Earthquake Time Bombs

In a media interview in January 2010, scientist Robert Yeats sounded the alarm on Port-au-Prince, Haiti, as an "earthquake time-bomb," a region at critical risk of major seismic activity. One week later, an earthquake of catastrophic proportions struck the city, leaving over 100,000 dead and triggering a humanitarian crisis. No one could have predicted the exact timing of the Haiti earthquake, but by analyzing its proximity to an active fault and its earthquake history, Yeats was able to point out the severity of the threat to Port-au-Prince. He forewarned that Haiti, the poorest country in the Western Hemisphere, was woefully unprepared for such a major quake.

Now, in a shocking and timely study, Yeats sheds new light on other earthquake hot-spots around the world and the communities at risk, including Caracas, Kabul, Tehran, and Jerusalem. He examines these seismic threats in the context of recent cultural history, including economic development, national politics, and international conflicts, and draws comparisons between the capacity of first-world and developing-world countries to prepare for the inevitable. The killer combination of mass migration to megacities coupled with poor building standards is explored, while descriptions of emerging seismic resilience plans from some cities around the world provide a more hopeful picture.

Earthquake Time Bombs is essential reading for policy-makers, infrastructure and emergency planners, scientists, students, and anyone living in the shadow of an earthquake. This book raises the alarm so that we can protect our vulnerable cities ... before it's too late.

ROBERT YEATS is a Fellow of the American Association for the Advancement of Science and the Geological Society of America. He is senior consultant and partner in Earth Consultants International, an international firm focusing on earthquake hazards, and also an Emeritus Professor at Oregon State University, where an endowed professorship has been named in his honor. He has decades of experience in earthquake geology worldwide, including acting as chair of an active fault working group of the International Lithosphere Program for several years and writing four previous books: *Geology of Earthquakes* (with Kerry Sieh and Clarence R. Allen), *Living with Earthquakes in California*, *Living with Earthquakes in the Pacific Northwest*, and *Active Faults of the World*.

Earthquake Time Bombs

ROBERT YEATS

Oregon State University

CAMBRIDGE
UNIVERSITY PRESS

University Printing House, Cambridge CB2 8BS, United Kingdom

Cambridge University Press is part of the University of Cambridge.

It furthers the University's mission by disseminating knowledge in the pursuit of education, learning, and research at the highest international levels of excellence.

www.cambridge.org
Information on this title: www.cambridge.org/9781107085244

© Robert Yeats 2015

First published 2015

Printed in the United Kingdom by TJ International Ltd. Padstow Cornwall

A catalogue record for this publication is available from the British Library

Library of Congress Cataloguing in Publication data
Yeats, Robert S.
Earthquake time bombs / Robert Yeats, Oregon State University.
    pages   cm
ISBN 978-1-107-08524-4 (Hardback)
1. Earthquake prediction.   2. Earthquakes.   3. Seismology.   I. Title.
QE538.8.Y43 2015
551.22–dc23   2015017920

ISBN 978-1-107-08524-4 Hardback

Contents

Acknowledgments

I decided to write *Earthquake Time Bombs* after completing a book called *Active Faults of the World* for professionals, also published by Cambridge University Press. Most earthquake professionals know where the time bombs are located, and I decided that I needed to help them make this information available to the general public we all serve. We can say *where*, even if we can't say *when*.

Over the last few decades of my career, I have been in a position to interact with my counterparts worldwide, where I learned that they, too, know where if not when. I visited many of the time bomb faults in the field, and I was able to talk to my colleagues working on these faults as well as private citizens who would be affected by the next damaging earthquake. They freely shared their information, and many of them are referenced below and in individual time bomb chapters. These people around the world have been my teachers, and many have become lifelong friends. My hope, shared by them, is that the information contained in the book will save lives. The book is dedicated to them.

Finally, I owe a great debt to my editors at Cambridge, starting with Susan Francis, whom I have met on several occasions as she shepherded me through *Active Faults of the World*, and Zoë Pruce, whom I have never met, but through our emails and her encouragement, I know her well enough to say thanks, not just for her editing skills but for her patience.

Why this book?

At the beginning of 2010, I was interviewed by *Scientific American* for a New Year's article on earthquakes. The interviewer, Katie Harmon, was well informed, and I enjoyed talking with her. I had just finished the Caribbean chapter of a book I had been writing for Cambridge University Press, *Active Faults of the World*, a chapter that had been fascinating to work on because Caribbean earthquake faults are so close to the United States, including my birthplace, Miami, Florida.

Somehow, the conversation turned to Port-au-Prince, the capital of Haiti, the poorest country in the Western Hemisphere. People have been moving to Port-au-Prince by the hundreds of thousands because that is where the jobs are, or appear to be. But the city lacks any social services, and there is no plan for where the new arrivals will live. So they are packed into the most abysmal slums, with shanties that are so fragile that it appeared a slight breeze would blow them down.

Port-au-Prince has a huge problem that is unrecognized by the government: It is adjacent to the Enriquillo plate-boundary earthquake fault, a structure that had not sustained a major earthquake near the city since the middle of the eighteenth century, prior to American (and Haitian) independence. I pointed out to Katie Harmon that Port-au-Prince is a time bomb. At the time of the previous earthquake, the city was a small town, but now it has a population in the millions, many living in dilapidated housing. If the Enriquillo fault ruptured, it would be a disaster of monumental proportions.

One week later, Port-au-Prince was destroyed by a large earthquake, killing more than 100,000 of its inhabitants.

Needless to say, Katie now had a major story, and she was on the phone immediately, asking breathlessly, "Did you predict the earthquake?"

"No, no, not at all," I replied. In my book, I could point out *where* the dangerous faults are, and even how active they are, but I could not say *when* the next big earthquake would strike – tomorrow or a hundred years from now.

In fact, some of my colleagues who were doing research in the region actually secured an audience with Haitian officials high in the government, about the earthquake danger prior to the earthquake. They described to the president's senior advisers an eighteenth-century earthquake of magnitude 7.5 on the Enriquillo fault and pointed out that a repeat of this earthquake today could cost hundreds of thousands of lives because of the great increase in population since the previous earthquake.

The government officials listened with concern, then asked the Big Question: "When will the next earthquake strike our city?" The response of their visiting seismic experts spoke to the uncertainty of earthquake science. We can point out the danger areas, but we can't predict when the next Big One will strike. It could strike tomorrow or a century from now.

So what did the Haitian government do?

Nothing.

Before condemning Haiti's short-sightedness, consider that Haiti, in addition to being the poorest country in the Western Hemisphere, has major problems that need to be addressed today: hurricanes, epidemics, the influx of thousands of unemployed Haitians to the capital city, a decrepit building stock, and virtually no social services. So can we blame the president if he puts on the back burner an earthquake that might not strike in a hundred years, long after he has left office?

The failure to predict earthquakes is our failure as scientists, despite the fact that major earthquake programs in Japan, China, and the United States were established with prediction as one of their most important goals. It's the one big question society wants us to answer: "When's the next Big One?"

Some leading seismologists state that the structure of the Earth is too complicated, and we know too little about it to make earthquake prediction a practical near-term goal. So we focus on preparedness, asking people to prepare for a disaster that might not strike during their lifetimes.

However, we can say *where* even if we can't say *when*.

In addition, there are other earthquake time bombs, places where great numbers of people are moving to megacities, the greatest worldwide migration in human history. Many of these megacities are in poor countries lacking a social infrastructure, and several of these megacities are close to active faults. Some have experienced earthquakes in the past, but the cities were much smaller then than they are now, so earthquake losses were moderately low. Some of the time bombs are in the United States.

Once I realized this, I re-directed the focus *of Active Faults of the World*, the book I was writing, to call attention to these time bombs. For example, Kabul, Afghanistan, is a city that has been battered by decades of war. After the Soviets were expelled, millions of Afghan refugees returned to Kabul from camps in Pakistan and Iran and from the Afghan countryside. They had been terrorized by insurgent groups, most recently by the Taliban. When I was in Kabul in 2002, I found families living in the ruins of buildings scarred by war, buildings so fragile that it seemed that even a moderate earthquake would cause them to collapse. Attempts by my company and by the US Geological Survey (USGS) to address Kabul's earthquake hazard through USAID went nowhere, even after I pointed out that the nearby Chaman plate-boundary fault had generated an earthquake of magnitude 7.3 in 1505, and a repeat performance could kill more of Kabul's inhabitants than four decades of war.

Tehran is a desert city built at the southern edge of the snow-capped Alborz Mountains to take advantage of the streams coursing down their southern slopes. This desert city is now the capital of a great nation undergoing drastic change after the Islamic revolution of 1979. The population has swelled to greater than ten million. The Alborz Mountains near Tehran have not experienced an earthquake since 1830, but these mountains have been convulsed by earthquakes of magnitude 7.3 or larger in the past 1500 years. Just north of the city, the North Tehran thrust is active, and faults have been found by Iranian geologists working with foreign experts even within the densely urbanized Tehran plain. The hazard from other faults within the city itself is unknown.

Many Iranian geologists, seismologists, and engineers have a clear-eyed view of the earthquake threat to the capital, but the construction industry and the city's building inspectors are notoriously corrupt, and the likelihood that new apartment buildings to accommodate the increasing population will be built to modern building codes is relatively low. The corruption problem was demonstrated in the easternmost suburbs of Istanbul, Turkey, in the Izmit earthquake of magnitude 7.4 in 1999 on the North Anatolian fault. Turkey has world-class seismologists, earthquake geologists, and engineers, but these experts were unable to prevent the construction of shoddy apartment buildings to accommodate newly arrived Turks from the countryside looking for work in Istanbul. Thousands died.

Worse yet, the North Anatolian fault extends westward into the Sea of Marmara offshore from Istanbul, a city of 15 million. The city is under alert for an earthquake greater than magnitude 7 in a part of the fault lacking in earthquakes comparable to the gap filled by the last earthquake of magnitude 7.1 in 1766. The fault east and west of this part of the fault generated earthquakes earlier in the twentieth century, but this part has not. The city of Istanbul has finally awakened to the earthquake wake-up call from 1999, and the future for that city looks brighter because of the steps now being taken.

The list goes on: Caracas, Venezuela, close to a plate-boundary fault that extends along the north coast, probably generating an earthquake in 1812 that took the lives of 5–10% of the population of

Venezuela at that time; Guantánamo, Cuba, the large city and the US naval base and prison of the same name, close to the seismically active Oriente fault offshore to the south; Dhaka, Bangladesh, Chandigarh, India, and Islamabad-Rawalpindi, Pakistan, close to the seismically active Himalaya to the north; Nairobi, Kenya, close to the Laikipia Escarpment on the East African rift valley, generating the Subukia earthquake of magnitude 6.9 in 1928; and the cities of the central valley of Burma (Myanmar), including Yangon, Mandalay, and the new capital of Naypyidaw, all close to the plate-boundary Sagaing fault.

Not all the time bombs are in the developing world. The Los Angeles, California, metropolitan area has earthquake-protective laws in place, but engineers point out that a large number of nonductile concrete buildings are likely to fail the test of the next large earthquake. California's laws may be inadequate to protect its citizens from a large earthquake or to fend off developers who want to take the chance that they can build their project with a life of no more than 200 years and it won't be destroyed by an earthquake with a recurrence interval measured in thousands of years. The general public is aware that California is earthquake country, but not disturbed enough to strengthen buildings so that their inhabitants will survive the next inevitable earthquake. The Mayor of Los Angeles, assisted by the USGS, has taken on the challenge of making Los Angeles more resilient to a major earthquake on the southern San Andreas fault.

WILL THE NEXT GREAT URBAN EARTHQUAKE STRIKE ONE OF THESE CITIES?

Governments and societies tend to respond to a disaster only after it has happened, as illustrated by two hurricanes in the eastern United States, Katrina in 2005 and Sandy in 2012, and by the Kobe, Japan, earthquake of 1995. Society would also respond if scientists could predict the time, place, and magnitude of the next earthquake. Sadly, we cannot. However, some cities are developing plans to ensure their resilience following a major earthquake, an idea that I hope is a good sign for the future.

The purpose of this book is to describe the threat faced by many cities throughout the world in hopes that their societies will respond to the danger before the earthquake, rather than picking up the pieces after the disaster has occurred and thousands of victims have been buried. I call these cities *earthquake time bombs*.

The Cascadia subduction zone, where I live, contains several large cities that I regard as time bombs. Groups of experts have prepared resilience surveys in the states of Washington and Oregon to determine the consequences of *not* getting ready for the earthquake disaster ahead of time. This can be rephrased as the cost of doing nothing. Although becoming earthquake-ready is expensive, in the billions of dollars, the cost of doing nothing is worse because it affects the entire society, including loss of jobs and tax revenues, as well as the shutdown of essential services and lifelines for months or years. This problem is dealt with in an international project called Global Earthquake Model (GEM).

This book is a wake-up call to our society and its leaders. The next great earthquake will be a disaster, but failing to prepare for it will lead to a catastrophe.

The book is written in two parts, plus a conclusion, and I have presented basic material in such a way that you don't have to start at the beginning. The first part provides a background in earthquakes and plate tectonics, including the concept of geologic time and an explanation of why we, as scientists, cannot tell you when the next huge earthquake will strike, or where. You can use this first part as a reference. The second part describes several earthquake time bombs around the world, most of which you have heard of for reasons other than earthquakes, such as Caracas, Tehran, Jerusalem, or Kabul. Some of these time bombs are in unexpected places: Seattle, Los Angeles, Tokyo. Each of the time bomb chapters may be read on their own without going back to the explanatory Part I, although you may want to read the explanations as well as the descriptions of individual time bombs. I have provided references in case you want to learn more about a specific time bomb of particular concern to you.

Since three of the time bombs I describe are in the Pacific Northwest, I have a personal stake in this because I live here.

PART I Earthquakes, deep time, and the population explosion

1   Plate tectonics and why we have earthquakes

INTRODUCTION

During my lifetime, the earth sciences have undergone a scientific revolution, as significant as the discovery that matter is composed of atoms containing a nucleus and electrons, or the discovery of DNA. This is the theory of *plate tectonics*, which completely knocks the props out from what I was taught in graduate school. This theory provides an explanation of why Earth, probably unique among the inner planets, is afflicted with earthquakes. I lived through this paradigm shift, and I now regard it as one of the most thrilling scientific adventures in my own life. Some of the major discoveries in this revolution were made by graduate students who were my own age.

THE PARADIGM SHIFT TO PLATE TECTONICS

Up until World War II, most geologists had no doubt that continents and ocean basins had always been where we now find them, although Alfred Wegener, a German meteorologist, had postulated the theory of *continental drift* as early as 1912 in the first edition of his book, *The Origin of Continents and Oceans*. Wegener had observed that you could close up the Atlantic Ocean, and the continents would come back together, like fitting together the pieces of a jigsaw puzzle. In addition, after fitting the puzzle pieces back together into a supercontinent Wegener called *Pangaea*, the picture on top of the jigsaw puzzle matched. Fossil plants and shallow-water animals on one side of the puzzle were found to be similar to those on the other side, even though the two sides are now separated by thousands of miles of ocean floor. Wegener argued that these plants and animals could not have made their way from one continent across the deep ocean to the other.

Wegener's idea was rejected by the scientific establishment of Europe and North America, but it was attractive to geologists in the Union of South Africa, especially Alexander du Toit. He pointed out that the geology of mountain ranges in South Africa is very similar to the geology of mountain ranges in South America after the continents had been fitted back together.

As pointed out above, Wegener's theory was attacked by the most prominent scientists of the day, especially in the United States. The American Association of Petroleum Geologists organized a symposium in New York City in 1926 in which America's foremost geologists presented papers arguing against Wegener's theory of continental drift. Wegener gave a talk at this symposium, but the Americans weren't buying. They stated that the Earth's crust is simply too firm for the continents to "simply plow through." The correspondence of coastlines pointed out by Wegener had different explanations, and the similarity of fossils could be explained by "land bridges" between the now widely separated continents, although no clear evidence for such land bridges was ever advanced. Except for du Toit of South Africa, and his colleagues, the theory fell out of favor and was essentially abandoned.

While I was in graduate school in 1954–1958, all but one of my professors regarded the theory of continental drift as nonsense. The one professor who accepted the idea, Professor Harry Wheeler, was considered to be a bit loony. My fellow grad students used him to illustrate the point that there is a fine line separating genius from madness. Back to him later.

In 1912, Wegener himself, in the first edition of his book, had advanced the idea that answers the objections that would be raised later at the New York symposium. The floor of the Atlantic Ocean could be continually tearing apart, an idea that would later evolve into *sea-floor spreading*. However, the vehement opposition to continental drift as an explanation for the distribution of continents caused Wegener to abandon the idea of tearing apart the ocean floor in later editions of his book. Wegener's first love was exploration of the

Arctic, particularly the Greenland Icecap, and expeditions to that unknown region beginning in 1906 took priority over defending his continental drift theory. Four years after the New York symposium, in 1930, Wegener lost his life during his fourth Greenland expedition, when his team of explorers attempted to spend the winter on the icecap. His body is still buried there. But du Toit and a few others continued to support Wegener's idea of continental drift.

The debate foundered because of the lack of knowledge about the deep ocean floor, which was more poorly known than the surface of the moon. This lack of knowledge persisted despite the laying of trans-Atlantic cables in the mid-nineteenth century and the world-wide expedition of the HMS *Challenger* in 1872–1876.

This changed during World War II, when it became necessary for the Allies to map the topography of the ocean floor in their search for enemy submarines (a plot line used in the 1984 Tom Clancy novel and the movie, *The Hunt for Red October*). The US Navy developed echo sounders in which a ship would send a continuous sound signal (sonar) to the deep ocean floor, which was returned to the ship in such a way that the distance between the ocean surface and the ocean floor could be determined, and a profile of the ship's track could be plotted. (This is the same system of echo-location used by bats!)

During World War II, Professor Harry Hess of Princeton University was called to active duty, where he served as the commander of a Navy attack ship, the *Cape Johnson*. Hess used his authority to take his ship across sea-floor features that might provide insight into the origin of ocean basins, in addition to clues as to the hiding places of German submarines. The technology of echo-location improved so much that after the war it could be shown that, unlike the assumptions of the American participants at the New York symposium, the ocean floor is not a gentle landscape receiving the sediment of billions of years. It is, instead, highly complex, with undersea mountains and canyons, as shown in a new map of the Atlantic Ocean floor published after the war by Bruce Heezen and Marie Tharp of Columbia University. The most important of these mountains is the

Mid-Atlantic Ridge, which lies midway between the Americas and Africa, the longest mountain range on Earth (Figure 1.1). Another set of features, pointed out by Hess, was *deep-sea trenches,* linear deepwater zones adjacent to mountain ranges like the Andes and Aleutians (Figure 1.1). These trenches are the location of the ocean's greatest depths.

This led to a new age of discovery: a series of campaigns by major American oceanographic institutions to map the topography of the ocean floor, leading to the discovery of other mid-ocean ridges in the Indian and Pacific oceans, deep-sea trenches off the mountainous coasts of South and Central America, Mexico, Alaska, and Japan, and great faults that displace the mid-ocean ridges. One of these ridges, the Juan de Fuca Ridge, lies off the coast of the Pacific Northwest. At about this time, I realized that most of the tectonic concepts I had learned in graduate school (and had been questioned on during my PhD oral exams) were wrong. And I gained a new respect for that

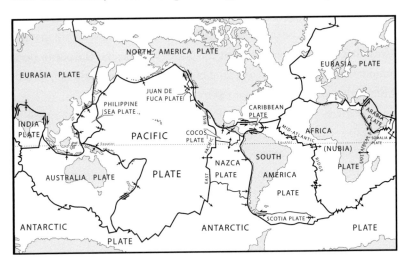

FIGURE 1.1 Division of Earth's surface into tectonic plates, all moving with respect to one another; microplates not shown. Diverging arrows, sea-floor spreading centers; converging arrows, subduction zones, including converging ranges (Himalaya, Alps); arrows parallel to boundary indicate strike slip, like the San Andreas fault. Source: image prepared by Kristi Weber.

"crackpot" Professor Harry Wheeler, who had not been a member of my PhD examining committee. Professor Wheeler did have some weird ideas, but on continental drift he was correct and far ahead of his time.

This led to a great scientific breakthrough by geologists in the United Kingdom. Professor Arthur Holmes (1978) of the University of Edinburgh wrote a textbook, *Principles of Physical Geology*, in which he used the known rates of radioactive decay of isotopes to argue for an Earth that is billions of years old. This led him to revive Wegener's theory of continental drift with his theory of convection cells (Figure 1.2), in which the ocean crust and underlying mantle could move by horizontal convection, thereby dissipating the Earth's internal heat. Because he had already pointed out the Earth's extreme age, the convection cells could move very slowly, and continents could be transported on top of them as passengers, like ice floes in the Arctic Ocean.

PALEOMAGNETISM

At the same time, geophysicists in Great Britain were measuring the magnetic properties of minerals. Our magnetic compasses depend on the flow of magnetic lines of force in the Earth that locate the magnetic north pole and permit ships to navigate on the open ocean, out of sight of land, when the stars are not visible. Navigators had been using magnetic lines of force to steer ships, even though no one knew why the magnetic lines of force exist. The magnetic north pole is not at the same place as the true north pole, the axis of Earth's rotation, and so maps were created showing the correction that your compass has to make to allow you to calculate the direction to true north. (My Boy Scout handbook had one of these maps, and I needed to use it to earn the Map-Reading Merit Badge.)

British scientists then began to measure the orientation of magnetic lines of force in volcanic rocks (*paleomagnetism*), which gives not only the direction to the magnetic pole of rotation but also the distance: The lines of force in a rock close to the magnetic pole would

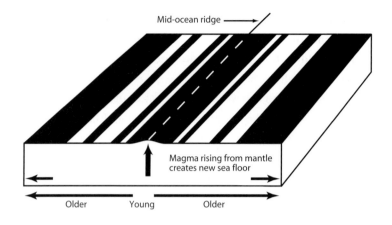

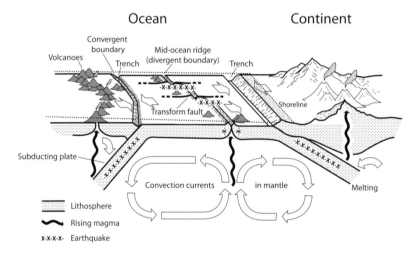

FIGURE 1.2 Top: mid-ocean ridge locates hot oceanic crust rising to the surface as older crust cools and moves away (diverging arrows). Spreading rate is dated by boundary between higher apparent magnetization of today's oceanic crust, magnetized in present magnetic field, (black; compass arrow points north) and lower apparent magnetization where the compass arrow points south (in white). This boundary is based on radiometric dating of volcanic rocks for which magnetization direction is known. Magnetic stripes make a symmetrical, mirror-image pattern. Bottom: relation of sea-floor spreading center, where oceanic crust is created, and subduction zone, where oceanic crust descends into Earth's mantle and is consumed, accompanied by great earthquakes (X). System driven by slowly moving convection currents in the mantle. Source: Ellen P. Metzger, Sea-Floor Spreading Teacher's Guide.

have a nearly vertical orientation whereas the lines of force in a rock close to the magnetic equator, one-quarter of the distance around the world, would be horizontal. These discoveries resulted in a startling discovery: The lines of force in volcanic rocks in some cases do not correspond to the lines of force of the present Earth's magnetic field. The lines of force point in other directions, and in some cases point south rather than north! (This was so bizarre that it was as if an instrument had just proven that the force of gravity in the geologic past had pointed up!) The conclusion that was reached was that the direction of the paleomagnetic lines of force could be used to tell you the latitude (north–south location with respect to the magnetic pole) the lava had been when it had cooled.

My employer at that time, Shell Oil Co., set up a paleomagnetics lab in Los Angeles to use the orientation of magnetic lines of force in drill cores to orient the cores with respect to north. This worked rather well except that some of the lines of force pointed south rather than north, exactly opposite the expected direction of the present day. Had we done something wrong? One theory was that the magnetic lines of force underwent self-reversal, changing their orientation for some unknown reason. In orienting cores, we learned to compensate by just assuming that a southward orientation could be used in the same way as the expected northward orientation. However, the lack of a rational scientific explanation was troubling.

Three young geologists at the US Geological Survey (USGS) and Stanford University collected a suite of volcanic rocks of various ages and measured their ages based on the radioactive decay of isotopes of the elements potassium and argon, ^{40}K to ^{40}Ar, following up on Arthur Holmes' discoveries. When their samples were taken from the outcrop, they were oriented in the present Earth's field, and this enabled them to determine the orientation of the magnetic lines of force in each of the samples they had dated.

The magnetic lines of force in all the volcanic rocks younger than 780,000 years behaved as one would expect – that is, they pointed north, toward the present Earth's magnetic pole. On the other hand, to the great surprise of the paleomagnetics team, the lavas dated as 780,000 to 2,600,000 years old were characterized by magnetic lines of force that pointed south! One of the three investigators, Brent Dalrymple, who is about my age, moved to Oregon State University late in his career as the dean of the College of Oceanography and became my neighbor and friend. It was obvious that this fundamental discovery had been a landmark of his career and life.

This age relationship proved to be true worldwide, meaning that it did not matter where the volcanic rocks had cooled, the only factor affecting whether they were normally magnetized or reversely magnetized was their age. This worldwide relationship meant that a geologic time scale could be established based on paleomagnetic reversals rather than on fossils. The youngest – normally magnetized – rocks are referred to the Brunhes Normal Magnetic Chron, and the older rocks – where the magnetic lines of force point south – are referred to the Matuyama Reversed Magnetic Chron, both named for pioneer investigators in the new field of paleomagnetism. There was more to come: The next epoch back in time is the Gauss Normal Magnetic Chron. In addition, there are shorter-term changes: the Jaramillo Normal Subchron in the upper part of the Matuyama and the Olduvai Normal Subchron in its lower part.

This showed that the change from normal to reverse was not part of a regular cycle like the tides or the phases of the moon. It was irregular, more like the timing of someone entering a room and turning on a light switch and someone else later turning the light switch off. Other workers extended the magnetic time scale back more than 150 million years. The conclusion from this was that in dating volcanic rocks, the Earth acts as its own magnetic tape recorder!

This changed the focus of oceanographic expeditions, and the next discoveries were made at sea in the early 1960s. Magnetic surveys over the Mid-Atlantic Ridge and the Juan de Fuca Ridge showed that the high- and low-intensity anomalies are quite linear and are parallel to the Mid-Atlantic and Juan de Fuca Ridges. North of the Equator, normally magnetized ocean crust at the mid-ocean ridges shows a more intense magnetic signal than reversely magnet-ized crust, which tends to diminish and cancel out the strength of part of the signal received by the ship's magnetometer. Two British geo-physicists, Fred Vine and Drummond Matthews, showed that not only were these signals linear, they also divided into mirror images, separated by the ridge crest (Figure 1.2). (When I heard their paper at a professional meeting, I realized that I had to unlearn all the "facts" about tectonics I had been taught in graduate school.)

The irregularity of ages of chrons and subchrons was key to the next step in identifying them on and near the mid-ocean ridges (Figure 1.2). The central anomaly is of high magnetic intensity; this had to have been formed during the Brunhes Chron. The lower-intensity anomalies on either side formed during the Matuyama Chron because the general magnetism of the Earth is partly cancelled out by the paleomagnetism of the oceanic crust. The Matuyama contains strips of higher-intensity crust that were correlated to the Jaramillo and Olduvai normal subchrons. Not only did this discovery prove sea-floor spreading, the rate at which spreading took place could be determined because Dalrymple and his colleagues had determined the ages of the magnetic chron boundaries.

These discoveries did not come easily. An early proponent of this view, Lawrence Morley of Canada, had submitted a paper advo-cating this concept, but his paper was rejected by all the major scien-tific journals. Alfred Wegener would have understood.

SEA-FLOOR SPREADING

The next step was to verify that the magnetic boundaries measured at sea actually correspond to the age of the oldest sediments overlying ocean crust. This required taking core samples on the flanks of the mid-ocean ridges which should show increasing ages with depth based on microscopic fossils in the sediments, with the sediment just above ocean basalt only slightly younger than the age of the underlying basalt, as predicted by the magnetometer readings. The oceanic crust beneath the oldest sediment recovered in the cores would have been at the mid-ocean ridge when it cooled. The oceanic crust should have been overlain directly by deep-sea sediments of the same age or slightly younger (Figure 1.2).

Proposals were submitted to the National Science Foundation to use a converted oil-industry drilling vessel, the *Glomar Challenger*, to test the theory of *sea-floor spreading* by collecting drill cores from the flanks of the two best-known mid-ocean ridges, the Mid-Atlantic Ridge and the East Pacific Rise. According to the new theory, the farther away we drilled from the crest of the East Pacific Rise, the older the basal sediments above basalt should be, transported away from the ridge crest by sea-floor spreading.

I was privileged to be a scientist aboard a cruise of the *Glomar Challenger* in 1971, in which we drilled a set of core holes down the western flank of the East Pacific Rise between South America and Hawaii. Paleontologists aboard ship studied samples of sediment from the cores under the microscope and identified microfossils for which the ages were known. The most exciting sediment cores were those just above ocean basalt. In every case, the age of these oldest sediments agreed with the age predicted by the magnetic anomaly at the site, which had been measured by a magnetometer survey before our voyage. We transmitted the ages by radio to headquarters, then at Scripps Institution of Oceanography, showing not only the agreement with the magnetometer ages, but calculating the sea-floor spreading rate from our results. Our scientific crew, and scientists from other cruises, had confirmed the concept of sea-floor spreading. It was an amazing time to be a marine geologist!

TRANSFORM FAULTS

The topographic mapping with echo-sounders showed that mid-ocean ridges are offset from place to place by great faults. In 1965, a Canadian geophysicist, J. Tuzo Wilson of the University of Toronto, showed that these faults were different from faults that offset a previously continuous feature. Returning to Wegener's jigsaw puzzle analogy, they were similar to the sides of tabs on a jigsaw puzzle piece. When you slowly pull the puzzle apart at these tabs, the ends of the tabs open up (spreading centers), and the sides slide past each other, like faults (Figure 1.3).

Because these are not ordinary faults, Wilson called them *transform faults* where they also formed the plate boundary. Some of the

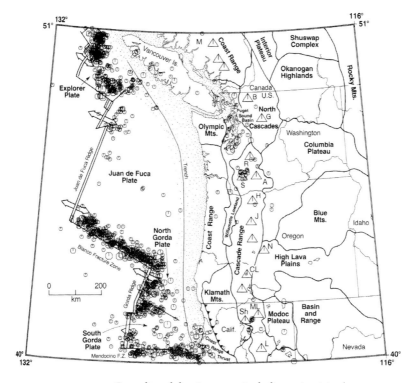

FIGURE 1.3 Cascadia subduction zone, including seismicity (open symbols, larger symbols for larger earthquakes). Open triangles: volcanoes in Cascade Volcanic Arc, identified by letters. Double lines with arrows pointing outward: sea-floor spreading centers. Note absence of seismicity at the subduction zone itself (labeled trench). Pacific plate is west of major spreading center; other plates to the east are identified. Most active seismically are Explorer and Gorda plates and Blanco and Mendocino fracture zones. In a subduction-zone earthquake, communities on the coast could be cut off from Willamette Lowland and Puget Sound for as long as several years. Source: Pacific Northwest Seismic Network and Yeats (2012, Fig. 2.15a).

transform faults on the Pacific Ocean floor extend westward from North America and turn into ordinary faults in the Pacific plate for more than 2500 miles (4000 km). The longest is the Mendocino fault (Figure 1.3), which intersects the north coast of California south of Eureka, near Cape Mendocino. There, the transform fault turns to the southeast and becomes the San Andreas earthquake fault, a transform fault that extends through the city of San Francisco. Professor Tanya Atwater of the University of California at Santa Barbara, while still a graduate student at Scripps, made the correlation between the off-shore Mendocino transform fault and the onshore San Andreas fault. Off the coast of the Pacific Northwest, another transform fault, the Blanco Fracture Zone, connects two sea-floor spreading centers, the Juan de Fuca Ridge and the smaller Gorda Ridge (Figure 1.3).

TRENCHES AND SUBDUCTION ZONES

Sea-floor spreading had been demonstrated, but it raised a question because new crust is formed at spreading centers. Does this mean that we live on an expanding Earth in which the area of the Earth is increasing, as some scientists were suggesting? *Trenches* were already known to be the most extreme examples of deep-sea topography on Earth (Figure 1.2). The Challenger Deep in the Marianas Trench is almost seven miles (10.9 km) deep, and the Mindanao Deep in the Philippine Trench is a close second in depth. In addition, most trenches are flanked by active volcanoes, such as Mount St. Helens and Mount Rainier in Washington State, Mt. Fuji in Japan, and Kraka-toa between Java and Sumatra. But what is the origin of these extreme features? One view was that trenches were the surface expression of great vertical faults.

The answer came from the Earth itself in the form of two super-quakes, one in southern Chile in 1960, of magnitude 9.5, and the other in the Gulf of Alaska in 1964, of magnitude 9.2 (Figure 1.4). Technology helped here, because in the early 1960s Western nations had established the Worldwide Standardized Seismic Network (WWSSN) to locate explosions that might be caused by underground testing of nuclear weapons by the Soviet Union. Fortuitously, the network also recorded earthquakes. This network was up and running at the time of the Gulf of Alaska earthquake, and everything changed.

The Aleutian Trench marks not a vertical fault, but a fault that dips beneath the North American continent. Study of the earthquake signal measured on seismographs showed that the Pacific Ocean crust was driving northward against and beneath the continental crust of Alaska, a process that came to be known as *subduction*. Crust is created at mid-ocean ridges, but it is consumed by subduction at trenches. The Earth's subduction zones were described by two seismologists, Kiyoo Wadati of the Japan Meteorological Service and Hugo Benioff of Caltech, and the oceanic crust descending beneath the continents became known as Wadati–Benioff zones. Because the length of the trenches is not too different from the length of mid-ocean ridges, one could throw away the idea of an expanding Earth.

This discovery led to a new age of exploration, in this case a delineation of the depths of subduction zones worldwide. Subduction zones were found descending beneath Japan, the Andes of South America, the Philippines, Java, Sumatra, New Zealand, and the Pacific Northwest (the last illustrated in the map in the lower right corner of Figure 1.4). One of the subduction zones, off the South Pacific volcanic island nation of Tonga, was traced to a depth greater than 400 miles (660 km) beneath the ocean surface.

This provided an explanation of why the subduction zones are bounded on the continental side by volcanoes (marked by triangles in the Cascade Range in Figure 1.3). The subducting oceanic crust has a lot of water caught up in it, and the tendency of this water to reduce the melting temperature of the mantle rock leads to eruptions of volcanoes like Mt. St. Helens (S, Figure 1.3).

Subduction zones are extreme in one additional way: They are the sources of most of the seismic energy released on Earth, including the world's largest earthquakes (Figure 1.5). But these superquakes originate within a few tens of miles of the Earth's surface. Earthquakes that define the downgoing plate at greater depths on a

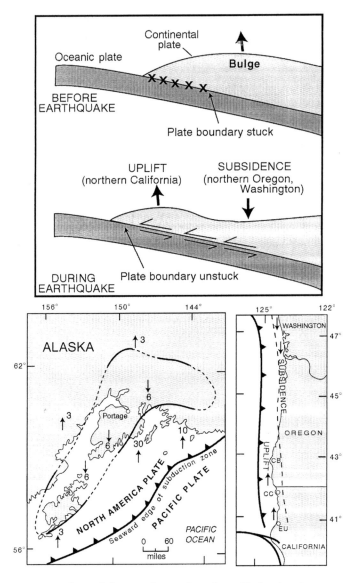

FIGURE 1.4 Lower left: map of coastal southern Alaska showing area affected by magnitude 9.2 earthquake in 1964. Areas of uplift (up arrows) and subsidence (down arrows) north of plate boundary (solid line with triangles in direction of continent). Upper diagram: cross-sections show build-up of strain in North American continental edge before the next subduction-zone earthquake. Plate boundary at trench is stuck (Xs),

subduction zone are, in general, not as large, although they tend to be felt at much greater distances than those near the surface.

As oceanic crust and continental crust are pushed together, the continental crust is forced downward, like an overweight diver standing on and bending a gigantic diving board (Figure 1.2, lower right diagram; Figure 1.4, upper diagram). Eventually, this bending exceeds the strength of the boundary between continental and oceanic crust at the deep-sea trench, causing the continental crust to pop upward, just as the diving board would do when the diver jumped off. This suddenly displaces the ocean water above the continental crust near the trench, acting like a giant paddle, producing an ocean wave called a *tsunami*. The tsunami generated by the March 2011 Tohoku-oki superquake of magnitude 9 off northeast Japan traveled across the Pacific Ocean at the speed of a commercial jetliner, causing big waves off the west coast of North America, thousands of miles away from the Japan earthquake. The tsunami produced by the 1964 Gulf of Alaska superquake traveled down the west coast of North America, causing losses of life in Oregon and California. I consider the tsunami hazard in greater detail later, in a separate chapter on the Cascadia subduction zone.

PLATE TECTONICS

The new worldwide seismic network enabled seismologists for the first time to construct a map of the Earth, both continents and ocean basins, that locates earthquakes. These earthquakes, concentrated at

---

Caption for figure 1.4 (*cont.*)
causing the trench to be dragged downward. Strain overcomes the stuck zone, producing a large earthquake. Sudden uplift of offshore edge of the continental plate forces water upward, generating a tsunami. Lower right: similar build-up of strain in Cascadia subduction zone. The next earthquake will be accompanied by uplift of coast south of Cape Blanco (CB). Farther north along Oregon and Washington coast, the earthquake will be accompanied by subsidence. Source: modified from Brian Atwater, USGS, and Yeats (2004, Fig. 4.11).

mid-ocean ridges, transform faults, and subduction zones, bound wide regions on continents and ocean basins with relatively few earthquakes (Figure 1.5). The earthquakes define the boundaries of tectonic plates, which are in slow motion with respect to adjacent plates, like gigantic ice floes. In some cases, plates move toward and beneath their neighbors, in other cases they move apart, and in still other cases they move sideways past one another, as transform faults like the San Andreas fault do (Figure 1.1).

One of the best-studied tectonic plates is the North American plate, which has as its eastern boundary the Mid-Atlantic Ridge and as its western boundary the San Andreas fault in California and the Cascadia subduction zone in the Pacific Northwest (Figure 1.1). The North American plate consists of the Atlantic oceanic crust west of the Mid-Atlantic Ridge and the continental crust of North America. The largest plate is the Pacific plate, which extends from the San Andreas fault and the Juan de Fuca Ridge west to great trenches off the coast of Japan and New Zealand (Figure 1.1). Smaller plates include the Juan de Fuca plate, which has its western boundary at the Juan de Fuca Ridge and its eastern boundary as the Cascadia subduction zone (Figures 1.3, 1.4).

The next step was to construct a map of the motions of all the plates with respect to all the other plates. Much of this was done by Tanya Atwater and her colleague Peter Molnar, now of the University of Colorado. They began their work with the observation that "As Columbus discovered, the Earth is round." They worked out the rates each plate was moving with respect to the other plates, and, sure enough, the rates matched around the world, another demonstration that the Earth was not expanding.

I have pointed out how sea-floor spreading allows the determination of how fast one plate is separating from its neighbor, but a more direct method was needed. One way is to work out the rates at which radio telescopes on Earth are moving with respect to one another based on signals from quasars from outer space, a technique known as very-long-baseline interferometry (VLBI). Another way, with wider

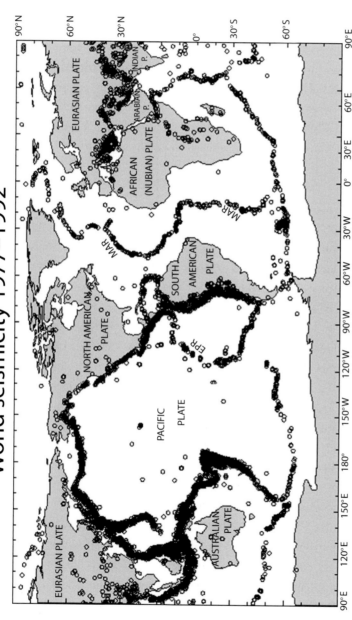

FIGURE 1.5 Map of the Earth showing earthquakes with magnitude greater than 5.5 from 1977 to 1992. Earthquakes follow narrow bands along the Mid-Atlantic Ridge (MAR) and East Pacific Rise (EPR), and a broader concentration of earthquakes at subduction zones, where most seismic energy is released. Areas in between, with few earthquakes, are tectonic plates, some of which are labeled. Not all earthquakes follow plate boundaries. Note earthquakes in western United States, western Europe, and China. Compare with Figure 1.1. Source: modified from Won-Young Kim, Lamont-Doherty Earth Observatory of Columbia University, © 1999.

application since there aren't many radio telescopes, is the Global Positioning System (GPS), which you use to locate yourself on a backpacking trip or in your car. These measurements are based on a "constellation" of space satellites that provide reference signals that can be compared with sites on land that can lock onto several satellites at the same time. Rates of offset at plate boundaries have been made for the few decades that GPS has been operational, which can be compared with measurements at spreading centers based on hundreds or thousands of years. GPS measurements below the ocean surface have been possible only rarely, for example, after the Tohoku-oki earthquake of March 2011 off northeastern Japan.

PLATE TECTONICS AND EARTHQUAKES

A worldwide map of tectonic plates shows that most of the earthquakes are found at plate boundaries (Figure 1.5). We know also that most of the seismic energy release takes place at subduction zones, with a lesser amount at transform faults and a still-lesser amount at spreading centers. The reason for this is that the strength of the Earth's crust is strongly affected by its temperature. New oceanic crust formed at spreading centers is hot and weak, and earthquakes produced at spreading centers are smaller than those at other types of plate boundaries. On the other hand, crust at subduction zones is likely to be older and colder, and this implies that it is much stronger. When strong, cold crust is subjected to the strain of subduction, it absorbs much more elastic strain by bending before rupturing in a great earthquake.

Does this mean that one can measure the rate at which, say, the Pacific plate is moving past the North American plate, and from that rate determine exactly when the next earthquake will strike? That would have solved the forecasting problem, but unfortunately it isn't that simple, particularly the "exactly" part. The reason is that the Earth's crust is complex, and its temperature is only one of the variables that might affect when it will rupture in the next earthquake. For example, the Cascadia subduction zone is known to have produced many subduction-zone earthquakes in the past 10,000 years, but seismicity at the Cascadia plate boundary today is close to zero (Figure 1.3). Slip rates on active faults in California based on GPS are in some cases similar to longer-term rates, and in other cases they are either slightly faster or slower.

This is the problem with using long-term rates between plates based on plate tectonics over geological time and rates based on active seismicity or GPS over only a few years. The Earth's crust is too complicated to permit a generalization that would allow us to say exactly when the next earthquake will strike. Plate-boundary rates permit the assignment of a greater hazard to, say, the Cascadia subduction zone or San Andreas fault in comparison with the east coast of the United States, where rates are extremely slow or unknown, but our knowledge of the local strength of the crust is insufficient to come up with an estimate of exactly when the next earthquake will strike.

As a result, the new knowledge of plate tectonics enables scientists to say *why* we have earthquakes, but science cannot say precisely *when*: tomorrow or a hundred years from now.

BOX 1.1 **The paradigm shift to plate tectonics: a personal adventure**

In graduate school, I was taught that mountains were built by earth deformation, including great faults, but my professors assumed that the Atlantic, Pacific, and other ocean basins were permanent and dated back to the early Earth. For me, the paradigm shift began with the San Andreas earthquake fault.

Before I started grad school, a paper had been published in 1953 by two geologists with Richfield Oil Corporation (later AtlanticRichfield or ARCO), stating that the opposite sides of the fault had moved horizontally hundreds of miles over a period of millions of years. This idea was ridiculed by most of my professors, except for that "nut case" Prof. Harry Wheeler, mentioned above. It was also condemned by the management of my employer, Shell Oil

BOX 1.1 **(cont.)**

Company, many of whom had been educated by the most famous structural geologist on the West Coast, Professor Andrew C. Lawson of the University of California at Berkeley. Lawson had directed a geological study of the San Andreas fault after the 1906 earthquake, and he recognized the evidence for strike slip, where the fault displacement was in a horizontal direction (Figure 1.6), but he did not believe that these offsets proved displacements of hundreds of miles. People who believed otherwise were shouted down.

At Shell, I became interested in Baja California, a long peninsula separated from the Mexican mainland by the Gulf of California. Professor Gordon Gastil of San Diego State University had been mapping the Baja peninsula, and he suggested that the peninsula had formerly nestled close to the rest of Mexico but was moved away by strike-slip on the San Andreas fault, opening up the Gulf of California. I must confess that when I read about this work, it gave me a queasy feeling, as if one of my basic scientific assumptions had been shredded.

Then I attended the national meeting of the Geological Society of America in San Francisco, the site of the 1906 earthquake, and I heard a series of papers by Fred Vine, Lynn Sykes, and other geophysicists who were advancing an idea they called the New Global Tectonics. I was sitting next to a friend from grad school, Dick Blank, and we said to each other that all that stuff we were taught in school can be tossed into the trash can. Harry Wheeler may have been a nut case, but he was right!

I then wrote an internal paper for Shell giving the implications of the New Global Tectonics to oil exploration, and the company asked me to give my findings to Shell offices around the country. Shell managers may not all have believed me, but they recognized that the ideas were worth presenting to other Shell exploration offices. At about the same time, excited by the new science I was learning, I accepted a teaching job at Ohio University. I was invited to give my talk at a special symposium on the San Andreas fault at Stanford

BOX 1.1 (cont.)

University, which required getting permission from Shell.
Surprisingly, permission was quickly granted, possibly because
upper management, still influenced by Andrew Lawson, didn't really
believe it, and didn't think that my giving the talk would give other
companies a competitive advantage! However, Shell provided me
with encouragement and gave me the chance to develop my ideas
using the new paradigm of plate tectonics.

At Ohio University, I had to teach a class in tectonics, the large-
scale deformation of continents, and mountain ranges. There were

FIGURE 1.6 Fence near Bolinas, Marin County, California, offset ten
feet by right-lateral strike slip on the San Andreas fault during the
1906 San Francisco earthquake. View northeast. Source: photo by
G. K. Gilbert, USGS. Photo from Karl Steinbrugge NISEE-PEER
collection at University of California Berkeley.

BOX 1.1 (cont.)

no textbooks, and the papers were only then being written. I wrote two of the papers myself, and I worried that someone out there would find the flaws in my logic and shoot me down. Fortunately for me, other geologists and geophysicists my age were going through the same process, and we talked to each other, over beers and other libations.

I have been a bit critical of several of my grad school professors at the University of Washington, but one of them, Professor J. Hoover Mackin, by his teaching methods, made my conversion possible. He taught us to respect the evidence and "think outside the box," although that phrase was not used in the 1960s. Mackin taught us to keep an open mind and not become tied to one explanation before considering all the evidence. Always think of all the ways you can to explain something. They might all be wrong, but one of the explanations might be closer to the truth, a pathway to the answer, and something that could be tested in the field. The method of multiple working hypotheses was not new; it was developed a century ago by a geology professor at the University of Wisconsin. Unfortunately, it was not used in the New York symposium that shot down Alfred Wegener's ideas about continental drift.

2 An earthquake primer

The preceding chapter shows why we have earthquakes. They are a product of plate tectonics, in which parts of the Earth's crust move with respect to adjacent parts due to slow convection in the underlying mantle (Figure 1.2, lower diagram). The present chapter presents a brief summary of how we describe earthquakes, including terms like *magnitude* and *intensity*, defined below.

An earthquake happens when two sections of crust, in some cases at a tectonic plate boundary, become so stressed that the strength of the rock is exceeded, generating an earthquake as the masses of rock on opposite sides of a zone of weakness suddenly shift past each other. Some part of the crust might be weaker than other parts, and so the earthquake breaks along this weaker zone. In this respect, the Earth behaves like a poorly constructed building, which is subject to collapse because of failure of its weakest part. This is likely to be a fault, a zone of weakness that has ruptured before, and ruptures again, accompanied by an earthquake.

The Earth's crust is *elastic*, that is, it bends when it is stressed, and the bend goes back to its original shape when the stress is removed. One example is the ocean crust near the Hawaiian Islands, which is bowed downward by the massive weight of Hawaiian volcanoes, the way your bed would behave if a very heavy individual lay down on it. When your heavy friend gets off the bed, it should return to its earlier shape, if it has good springs.

As two plates move toward each other in a subduction zone, the upper plate is slowly bowed downward elastically, like a diving board, until the strength of the subduction zone at the plate boundary is exceeded, producing an earthquake (Figure 1.4, upper diagram). When that happens to a subduction zone that is underwater, the upper plate that has been bowed downward snaps back. In this comparison, the Earth's crust is like a diving board, except that it is underwater. When the diver jumps off, the board rebounds to its position before the diver walked out on it. As the Earth's crust rebounds beneath the sea, it suddenly pushes upward all the seawater on top of it, producing a tsunami.

HOW BIG IS THE EARTHQUAKE, AND WHERE IS IT?

Before the last part of the nineteenth century, scientists rated the size of earthquakes based on the damage they did. But the invention of seismographs now permits a measure of the energy released by the earthquake. The seismic record shows a series of wiggles that represent the release of elastic energy, further evidence that the Earth's crust is elastic. Charles Richter of the United States and Kiyoo Wadati of Japan determined that the size of the wiggle is a measure of the energy released by the fault rupture as well as the distance between the earthquake and the seismograph.

The location of an earthquake within the Earth's crust is called its *focus* or *hypocenter*. The point on the Earth's surface directly above the hypocenter is the *epicenter* (Figure 2.1).

In 1935, Richter devised a scale of released energy called the *magnitude scale*, with the larger numbers referring to larger earthquakes. He calibrated his scale on the basis of offset of the wiggles of the needle of a particular kind of earthquake recorder in use during the 1930s, called the Wood-Anderson seismograph, at a distance of 62 miles (100 km) from the source of the earthquake. Richter never made claims that this would be a precise scale, or that it would apply to regions other than southern California. The numbers used, from less than 1 to about 9, are logarithmic, meaning that a magnitude 5 would be about ten times the size of a magnitude 4. In terms of energy released, the difference between two adjacent magnitude numbers is about 30 times. The Richter magnitude is now known as the *local*

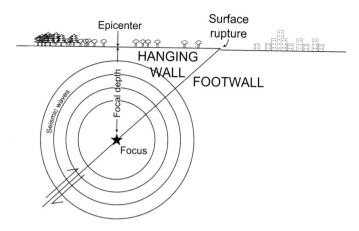

FIGURE 2.1 Cross-section of Earth's crust showing an earthquake fault, the point within the crust where the earthquake starts (focus or hypocenter) and the point on the Earth's surface directly above the focus (epicenter). Arrows show the relative displacement on the fault due to the earthquake. In some cases, the fault ruptures to the surface. Seismic waves extend out from the focus and produce the strong shaking experienced during an earthquake. Source: Yeats (2004, Fig. 3.4).

magnitude, M_L, although the news media still generally refer to it as the Richter magnitude (much to Richter's chagrin). You might not feel a magnitude less than 2. More recently, the *moment magnitude scale* has been introduced because the local magnitude scale could not express the energy released in superquakes. This scale, or M_w, uses the area of the surface rupture, the displacement along the fault surface, and a number for the strength of the fault. It has the advantage that it can be used for small and very large earthquakes, including the March 11, 2011 Tohoku-oki earthquake of M_w 9.

THE EARTHQUAKE SYMPHONY

I am a music lover, and I like to think about earthquakes as similar to a symphony orchestra ("the music of the spheres"). There are piccolos, flutes, and violins (high-frequency waves) and tubas, bass fiddles, and bass drums (low-frequency waves). Low-frequency waves have the advantage that they can be transmitted great distances, so that our seismograph in Corvallis, Oregon, was able to record the low-frequency waves of the Tohoku-oki earthquake off the coast of Japan. This is similar to a pedestrian listening to the boom box in a passing car. You can't make out the voices but you do hear the boom of the low-frequency basses, which is, I suppose, why they call it a boom box. The seismogram that records the wiggle of the needle is recording a mixture of waves of different loudness and different frequencies, just as your ear does when you hear the different frequencies and different loudness of musical instruments in an orchestra.

Earthquakes produce two very different kinds of waves, P-waves and S-waves (Figure 2.2). A P-wave can be visualized by thinking of when you play pool and your cue ball hits other balls on the table. The cue ball transfers its energy to other balls on the pool table by compressing them, and causes the other balls to move and possibly to go into a corner pocket. An S-wave is like flipping a clothes line tied to a pole, or the waves on the surface of a pond after you have thrown a pebble into it. The direction of motion is at right angles to the direction the wave is moving. P-waves travel through the Earth faster than S-waves, so you can use the difference in arrival times of P- and S-waves on your seismogram to tell you how far the earthquake is from your seismograph station.

The P-wave and S-wave are *body waves* that travel through the Earth, and the different speeds they travel and the way they lose their energy with distance (*attenuation*) teach the seismologist about the Earth materials the waves are passing through – a field called *seismic tomography*. The principle is similar to tomography used in studying the human body, in that sound waves pass through our body and provide images of what our insides look like.

A third type of wave produced by earthquakes is the *surface wave*, which travels along the surface of the Earth and can cause major damage. Many people who have experienced an earthquake report that they can see the ground moving up and down in front of them, like an ocean wave. Surface waves can produce a lot of damage

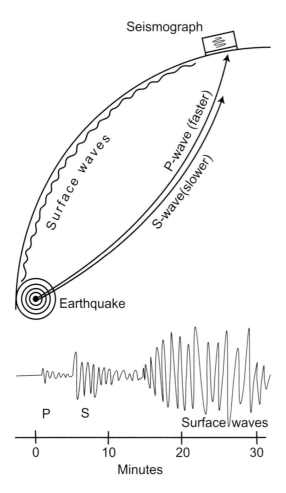

FIGURE 2.2 Top: cross-section of the Earth showing that an earthquake produces P (compressional) waves that travel faster and reach the seismograph first, and S (shear) waves, that are slower. Both are transmitted through the Earth and are called body waves. Even slower are surface waves that run along the surface of the earth and do much of the damage. The seismogram at the bottom reads from left to right, showing the successive arrivals of P-, S-, and surface waves. The length of time between arrival of the P-wave and S-wave is used to determine how far the earthquake is from the seismograph. Source: Yeats (2004, Fig. 3.11).

because the wave moving close to the surface can cause buildings to tilt, fall over, and collapse. Seismographs can measure the speed of the wave as well as its acceleration, which is the rate the wave causes objects it encounters to speed up, like accelerating an automobile from a stationary start to, say, 60 miles (100 km) per hour. If a seismic wave accelerates at a rate greater than Earth's gravity, objects may be thrown into the air.

Why does strong ground shaking continue after a fault ruptures at an earthquake's hypocenter? To return to our comparison of an earthquake to musical instruments in an orchestra, an earthquake is more like a pair of cymbals than the breaking of a stick. The cymbals clash and the sound continues for a period of time after the cymbals are struck. Strike a note on the piano, and it continues to make sound, although at a diminishing rate. The earthquake strikes a note, and because the Earth's crust is elastic, the note continues to shake the Earth, decreasing in its "loudness" with time.

The strong shaking continues in very large earthquakes because the fault rupture is propagating along the fault, like tearing a piece of paper. An earthquake off the northwest coast of Sumatra continued to rupture the subduction zone to the north for several minutes after the first shock. This is why subduction-zone earthquakes are so dangerous. A building might be able to withstand strong shaking for 60 seconds, but could not hold up if the strong shaking continued for several minutes.

The speed of an ongoing fault rupture, as opposed to the speed of a seismic wave passing through Earth's elastic crust, can now be measured using new, sophisticated seismic instruments. In some cases, the speed of a fault rupture has been observed directly at the surface. In October 1983, two elk hunters in Idaho observed surface rupture accompanying the Borah Peak earthquake. The propagating rupture zipped past the astonished hunters at a high rate, almost (but not quite) instantaneously.

EARTHQUAKE INTENSITY

Until the invention of the seismograph in the late nineteenth century, only earthquake intensity could be measured. But even since the invention of the seismograph, intensity measurements are still valuable because intensity measures the effects on buildings around us. If you feel an earthquake, you can contribute to science by going to the USGS website (http://earthquake.usgs.gov/earthquakes) and clicking on "Did You Feel It?" You will fill out a questionnaire that is used to work out the response of the Earth's crust to the earthquake, thereby making an important contribution to seismology.

In Western countries, we use the Modified Mercalli Intensity (MMI) scale, with intensity recorded in Roman numerals. An MMI intensity of III would be felt but would result in little or no damage, whereas MMI intensities of IX or X would cause major damage. Figure 2.3 is an example of an intensity map for a moderate-size earthquake in western Oregon. Countries in Central Asia and Japan use different intensity scales because their construction practices are very different from those in the United States or Western Europe. We don't have to worry about the collapse of mud-block buildings, which they do in places like Iran and Afghanistan.

Is it possible to calculate the magnitude of pre-instrumental earthquakes? Iran and Turkey, for example, have intensity data for earthquakes over a period of more than 2000 years. Seismologists work out the maximum intensity of an earthquake, generally (but not always) closest to its epicenter, and assign a magnitude by comparison with recent Iranian or Turkish earthquakes for which the magnitude has been established independently. Bill Bakun and Carl Wentworth (1997) of the USGS addressed this problem for pre-instrumental earthquakes by comparing the path between pre-instrumental earthquakes and sites where the damage had been established with similar paths of modern earthquakes with similar epicenters where the damage could be compared with the earlier earthquakes. They call magnitude determined in this way *intensity*

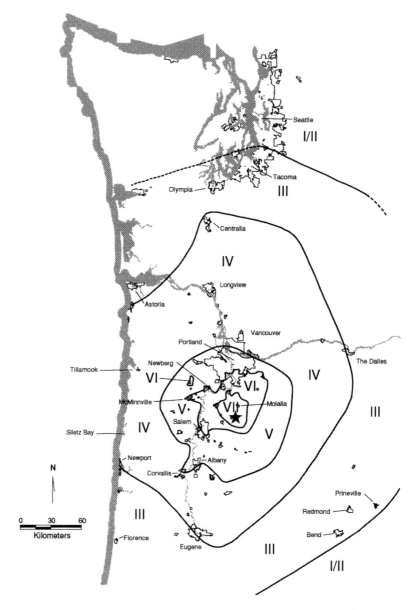

FIGURE 2.3 Modified Mercalli intensities (Roman numerals) of March 25, 1993, Scotts Mills earthquake in western Oregon. Star marks the epicenter determined from seismographs. The irregularity of the contours is due to differences in response of the Earth's crust in western Oregon due to shaking. Source: Gerald Black, Oregon Department of Geology and Mineral Industries, and Yeats (2004, Fig. 3.15).

magnitude, or M_i. This is an important thing to do because the recurrence time of most earthquakes on a given section of fault is longer than the time earthquakes have been measured using seismograms. Intensity magnitude has been useful in estimating the pre-seismograph magnitude of earthquakes in Middle Eastern countries as well as Japan. Chapter 15, on Istanbul, presents an example where the calculated magnitudes of pre-seismograph earthquakes were used to establish the present-day hazard in a city that had recently experienced a major damaging earthquake in 1999.

For more detailed information, see Bolt, (2004) and Richter, (1958).

3   **Deep time**

Chuck Denham and I were sitting in a bar in Mexico City at 9:30 a.m., planning a climb of Mt. Popocatépetl, the active volcano that looms over the eastern edge of the metropolis. Just as I was sipping my beer, the room started to sway, and glasses behind the bar started to tinkle and break, bringing forth a lot of swearing in Spanish. OMG, I said to Chuck, we're having an earthquake!

We then did a stupid thing and ran outside; stupid, because if bricks had started falling off the building, we could have been struck and killed. Once we were outside, it was hard to remain standing in the middle of the street. Parked cars were rocking back and forth, and buildings were clapping together like giant hands; fortunately for us, not shedding bricks onto the sidewalk. A siren wailed in the distance.

After what seemed forever, the shaking stopped and the earthquake was over. I looked at my watch and saw that the entire shaking episode had taken only 15 seconds! I could have sworn it was several times that long – more than a minute.

On March 11, 2011, my Oregon State University colleague, Chris Goldfinger, was in a conference north of Tokyo, Japan, discussing earthquakes, when the meeting room began to shake. Someone took a video of the confusion that descended upon the assembled scientists, again showing people rushing outside. In this case, the shaking kept going and going. When it stopped, Chris estimated that strong shaking had lasted five minutes, although other estimates were a minute or two less. But it seemed to people at the conference that the shaking would never stop.

The lesson in both cases was our perception of time, which seemed to almost stand still. The world could have been crashing around us, and we might lose our lives. To us, the entire experience was in slow motion, like an instant replay of football on television. In our minds, seconds became minutes.

This is one way that we experience an earthquake, as it is happening. But there is another way because the earthquake is a response to a build-up of strain in the Earth's crust. This strain build-up is slow, roughly the rate your fingernails grow, but it is inexorable, taking hundreds or even thousands of years. You can stare at your fingernails all you like, but you can't see them growing. You only know later that they have grown, after the fact, when you see that they need to be clipped.

Imagine putting a board in a huge vise, and slowly compressing the ends of the board, bending it into the shape of a bow. In geological terms, "slow" means infinitesimally slow, tectonically slow, so slow that you could not see the board bending, not even with time-lapse photography.

A number of years ago, my friend Joe Vance and I were climbing Mt. Olympus in Olympic National Park in Washington State. As we ascended, we could look over at the Blue Glacier Icefall, which some people use as a climbing route. The icefall was as still as could be, as stationary as Mount Olympus itself. A few weeks later, I looked at a video of the icefall taken with time-lapse photography, and it showed the icefall moving like a rockslide in slow motion. I thought at the time: no way could you get me to climb Mt. Olympus by way of the Blue Glacier! Yet in the Nepal Himalaya, the Khumbu Icefall is the standard climbing route from Everest Base Camp to the South Col, enroute to the summit of Mt. Everest. Yes, it could shatter and fail and sometimes does, with loss of life, but it is safe enough for climbing parties to go from Base Camp to the South Col on their way to the summit of Earth's highest peak. Part of the Khumbu Icefall did fail in 2014, resulting in the deaths of several Sherpa guides.

But what about the mountain itself? The great pyramid of Mt. Everest seems to be unmoving, even as ice tumbles from time to time down the Khumbu Icefall. Mt. Everest in photos I took in 1991 looks the same as photos taken by Sir Edmund Hillary and Tenzing Norgay in the 1953 first ascent. And yet Mount Everest *does* move, slowly rising between earthquakes as the Indian tectonic plate drives against and beneath Tibet. Over the centuries, Everest, along with other great peaks of the Himalaya, has been uplifted at the same time that erosion has worn these peaks down. This is true not only of the Himalaya but also of the Alps, the Sierra Nevada, the Rocky Mountains, the North Cascades, and even the Appalachians. The uplift is too slow to observe directly, but it can now be measured with highly sophisticated survey devices, some mounted on satellites and using GPS.

Not all the internal deformation of the Earth's crust is recorded as uplift. The San Andreas fault, the boundary between the Pacific and North American tectonic plates, can be seen from space as a straight-line scar across the California landscape, extending from Point Arena on the northern California coast, southeast past San Francisco to the Salton Sea in the southeastern corner of the state. It is not moving, and in the course of my own geological field work I have sometimes sat right on the fault and eaten my sack lunch. Some would call this tempting fate, but the possibility of the fault rupturing while I was sitting there having lunch is extremely small, and I feel quite safe. Roads, railways, and pipelines cross the fault, and so will you if you happen to drive your car from Los Angeles to Phoenix or Las Vegas. You might go from southern California to Las Vegas to gamble, but the gamble you take as you cross the San Andreas fault is very small compared with your gambling on slot machines or the roulette wheel.

Yet horizontal strain is building up across the fault because the San Andreas marks a plate boundary. The build-up is similar to compressing the board in the vise, slowly deforming it but doing so too slowly to see it happening. In the Coachella Valley of southeastern California, the fault has not had a rupture in more than 300 years, even though it has been building up strain that entire time. This strain build-up can be observed directly using time-lapse satellite imagery called InSAR.

Tectonic processes are so slow that we have to consider much longer spans of time to understand earthquakes on active faults. For

example, the Port-au-Prince, Haiti, earthquake in January 2010, ruptured part of an active plate-boundary fault. This fault had last ruptured 240 years earlier, in the middle of the eighteenth century, before
Haiti and the United States had gained their independence. After
that earthquake, both countries had revolutions, the French and
the British were kicked out, and the newly independent country of
Haiti developed without any thought that its new capital was close to
a plate-boundary fault that is prone to earthquakes. Between the
eighteenth-century earthquake and 2010, the plate-boundary fault
near Port-au-Prince, Haiti, was as quiet as could be. The people who
lived in Port-au-Prince and the rulers of Haiti had no idea that they
lived next to a tectonic time bomb.

Society comes to its own conclusions about where earthquake
country is, based on recent experience. When I lived in southern
California, my neighbors regarded San Francisco as the most dangerous place in the state because of the 1906 earthquake. The Los
Angeles basin had experienced a major earthquake in 1769 during
the Portolá exploratory expedition, and a mission on the coast was
damaged in a San Andreas fault earthquake in 1812. But these earthquakes struck too long ago to be remembered by Angelenos, and
so the dubious "honor" of being the earthquake capital of the state
went to San Francisco. In contrast, geologists are aware that the San
Andreas plate-boundary fault has produced many earthquakes over
thousands of years, and the 1906 San Francisco earthquake was only
the most recent. But the 1906 earthquake is remembered because it
was a major event, destroying the ninth largest city in the United
States during California's short human history.

In order to advise Californians about the hazard from the San
Andreas fault, earthquake geologists need to take a longer look,
working out its earthquake history over thousands of years. Most of
this history took place prior to California becoming a state, even prior
to its settlement by Native Americans more than 10,000 years ago.
This requires an investigation of its earthquake history based on geological evidence, a field known as *paleoseismology* (Figures 3.1, 3.2). If

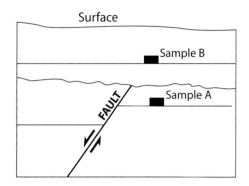

FIGURE 3.1 Radiocarbon dating a fault rupture in a backhoe trench by paleoseismology. Sample A dates the youngest sediments ruptured by the fault. Sample B is in sediments overlying the fault. The rupture cannot be dated directly except to say that it is younger than A and older than B.

FIGURE 3.2 Trench excavation of a strand of the La Rinconada fault east of the crest of the Andes in San Juan Province, Argentina. This fault may have been the source of an earthquake of moment magnitude 6.8 in 1952. The fault is present at depth, but it gives way upward to a fold and a gentle, partly eroded scarp at the surface. Backpack for scale. Photo by Carlos Costa, National University of San Luis, Argentina.

geologists could work out the earthquake history for thousands of years, they could say what the long-term slip rate on the fault is, even though the earthquakes are episodic and sudden, and are separated by long periods of inactivity, enabling me to have my lunch sitting on the fault and enabling you to travel from Los Angeles to Las Vegas without

worrying that my lunch or your trip might be interrupted by an earthquake. It could happen, but the chances of an earthquake striking just
when you or I were crossing the fault are very small.

If your house was built across the fault, that's another story
altogether.

Geologic time, like astronomical distances in light years, is
virtually impossible for us to imagine because we don't have anything
to relate it to. We can date rocks in the laboratory because we can
determine exactly how long it takes certain radioactive isotopes of
elements such as uranium or carbon to decay to other isotopes. From
this, we learn that the Earth is more than 4.5 billion years old, an
unimaginably long time. In dating past earthquakes, we don't need
to worry about most of that deep time, but we do have to take into
consideration materials that are up to a few tens of thousands of
years old.

But that's still too long a time for us to understand when it's
hard enough for us to visualize our own lifetime, less than 100 years.
When I was growing up, it seemed an impossibly long time until my
birthday or Christmas. As a small boy, I knew my grandfather, who
was born about the time of the Civil War, but he would have had to go
back to his own grandfather to have known an American alive
during the time of the previous earthquake striking Port-au-
Prince, Haiti.

Geologists work out the age of sediments that have been disturbed by earthquake faulting by the rate a radiogenic isotope of
carbon, ^{14}C, decays to non-radiogenic carbon, ^{12}C, a rate that is very
precisely known. This geologic clock relies on the fact that the carbon
in living things, including our bodies, contains a known amount of
^{14}C because we are constantly bombarded by cosmic rays from outer
space. When we die, our bodies stop accumulating ^{14}C, and our
^{14}C decays at a known rate. For organisms that died 100,000 years
ago, too little radiogenic ^{14}C is left to be measured. Fortunately, there
are other techniques in the geologist's bag of tricks, like optically
stimulated luminescence that requires that the sediment we want to

date must not have been exposed to sunlight since it was deposited. Deposits of volcanic ash, like the ash formed during the eruption of Mt. St. Helens in 1980, can be dated independently of the sediments enclosing the ash.

A backhoe or bulldozer trench across the trace of an active fault may be used to date an episode of faulting by dating the youngest sediments exposed in the walls of the trench that are offset by the fault and the oldest sediments that overlie the fault and are not offset (Figures 3.1, 3.2). For instance, the youngest sediments offset by the fault might be about 3000 years old, and the oldest sediments overlying the fault might be about 2000 years old. The geologist can then say that the earthquake that was accompanied by rupture on the fault is between 2000 and 3000 years old. I say "about" because radiocarbon dating is not accurate enough to get us back to the exact year an earthquake struck, as we will see when we discuss the date of the most recent earthquake on the Cascadia subduction zone. There are several reasons for this, including variation in the intensity of bombardment by cosmic rays of the materials we want to date. Incorporation of old, dead carbon or living carbon, like plant roots, in our sample would contaminate our sample and give us the wrong date.

The fault that ruptured in the examples above would be dated as *Holocene*, a term used for the past 11,700 years. A fault rupture of that age is said to be *active*, a term with legal significance in that in regions where legislation exists, such as California and parts of Utah, special considerations need to be taken before building near or across the fault. This is preceded by the *Pleistocene*, which dates back to 2,600,000 years and includes the ice ages. There are many older periods in the geological time scale, but these are the ones that matter the most to earthquake geologists. Together, the Pleistocene and Holocene make up the geologic period called the Quaternary.

The evidence for earthquakes older than the modern scientific age, where they may be measured by seismographs and other modern instruments, makes up the field of paleoseismology, which in a general way is geological, not seismographic, evidence for individual earthquakes. It is important to study earthquakes in this way because if we have the record of several earthquakes on a given segment of fault, we may have a better chance of saying when the next earthquake is likely to rupture that fault based on the slip rate on the fault.

A better chance, yes, but it's not a slam dunk.

Unfortunately, the Earth is too complicated for earthquakes to return at a precise time interval, like Halley's comet, even if we know the rate at which long-term slip is accumulating along that fault. Let's say that we know how fast strain is building up on a fault, and we also know the amount of slip that took place in the last earthquake. Can't we just divide the slip by the slip rate to give us the time between successive earthquakes?

Sadly, we cannot because the Earth's interior is too complex to lend itself to predicting exactly when a fault will break. Scientists have worked very hard at answering this question and predicting the time of the next earthquake, because the main thing you want to know from us is "when's the next earthquake?" And for the most part, we can't answer that question (see Chapter 4).

Despite our inability to give a precise date for the last, or the next, earthquake, paleoseismologists still learn as much as they can about active faults, especially their earthquake history. Specific parts of the North Anatolian fault in Turkey, a major hazard to the megacity of Istanbul, experienced more than one earthquake during the long recorded history of Turkey, lasting nearly 4000 years. For most faults, though, we need geological evidence for past earthquakes, so that we can say that earthquakes repeat on a given section of fault every few hundred years or every few thousand years. For the Cascadia subduction zone off the Pacific Northwest, we know of 19 previous great earthquakes, with the last one in AD 1700. But even with that information, including the knowledge that Cascadia will certainly have a huge earthquake sometime in the near future, we can't tell the people who live there (including my own family) whether the next earthquake will arrive tomorrow or a century from now.

Nonetheless, the knowledge that motion along an active fault takes place over thousands of years, reflecting the motion of plate boundaries, gives us some idea about the hazard that the fault poses to people and cities nearby. This makes it worthwhile for us to think in terms of a fault history lasting many thousands of years so that we can use the history of earthquakes along the fault as a guide to future behavior. That is why we must think in terms of long periods of time, even if we can't personally imagine these lengths of time.

Engineers and insurance actuaries would like to make forecasting earthquakes more precise because this would guide the insurance premiums they should charge or the strengthening of buildings they should require. As a result, a major research effort is underway to improve our ability to forecast earthquakes. So far, we have not had much success.

But we keep trying.

4   When's the next Big One?

INTRODUCTION AND FRUSTRATION

As earthquake scientists, we study a natural phenomenon that could destroy a city and could kill tens of thousands of people. For that reason, we are frequently interviewed on television and radio to discuss earthquakes. Invariably, the first thing our interviewer wants to know is "When's the next Big One?" When our answer is "I don't know, exactly," we frustrate our interviewer, who may start to lose interest. "You've been studying this problem for decades, and you *still* don't know?" But we are required to talk about identifying earthquakes in advance and say why we're not farther along in answering this question.

In our defense, it should be noted that it is difficult to predict *anything*. Will the stock market plunge us into a new recession like the one in 2008? Will we suffer a nuclear attack from North Korea? Will we suffer a drought next summer as bad as last summer? Each of these questions has its own army of prognosticators, so the earthquake people have lots of company.

The section on China (Chapter 12) describes the world's first successful earthquake "prediction" in Haicheng, Manchuria – an earthquake of magnitude 7.3 in February 1975. A year-and-a-half later, while the Chinese were still basking under the impression that they had figured out how to predict earthquakes, the nearby city of Tangshan was destroyed by an earthquake of magnitude 7.8, with the loss of more than 240,000 lives. The Haicheng earthquake had been preceded by foreshocks that had rattled the city and alerted its inhabitants to the danger, but the Tangshan earthquake had no foreshocks. It came as a complete surprise. Back to the drawing boards, not just the Chinese but all of us.

The US Geological Survey (USGS) thought it had a key to forecasting an earthquake on the San Andreas fault in a segment of the Coast Ranges of central California. This segment, at the tiny village of Parkfield, experienced moderate-sized earthquakes of magnitude 6 to 6.4 in 1857, 1881, 1901, 1922, 1934, and 1966, repeating every two to three decades. After the 1966 earthquake, the USGS installed a large array of instruments at Parkfield to record anomalies that might show signs of a forthcoming earthquake, calling this "a test of a hypothesis." Based on the regularity of moderate-sized earthquakes for the past century, the USGS thought the next Parkfield earthquake was likely to strike around 1988, 22 years after the preceding earthquake, give or take seven years. The California Office of Emergency Services got wind of this experiment and asked the USGS to make this forecast public, which was done. The year 1988 came and went; no earthquake. Finally, it arrived 18 years late, in 2004 – 38 years after the preceding earthquake. Again, back to the drawing boards.

The Parkfield experiment and its outcome could be compared with a man waiting for a bus that is due to arrive at noon. Noon comes and goes, no bus. The man thinks that the bus is late, and it will arrive any minute. His idea is that the likelihood of the bus arriving *increases* the longer he waits.

But maybe the bus had an accident, or a bridge was out. If this is so, then the likelihood of the bus arriving very soon *decreases* the longer the man waits. The "accident" affecting Parkfield might have been the May 2, 1983 Coalinga earthquake of magnitude 6.5 in the Coast Range east of Parkfield and east of the San Andreas fault. By altering the strain build-up in the Earth's crust, the Coalinga earthquake might have thrown the Parkfield earthquake off schedule. An earthquake of the expected size struck central California within the predicted time, but it was in the wrong place.

A successful earthquake prediction must include *time, place*, and *magnitude*. When the expected "Parkfield earthquake" finally arrived in 2004, it had place and magnitude about right, but it was 18 years late.

The problem with using this line of reasoning elsewhere is that the return time for earthquakes on most faults is not decades, but hundreds to thousands of years. If the man waiting for the bus is the governor of California, the uncertainty in our guess as to the time of arrival of the next earthquake on, say, the San Andreas fault in San Francisco could be plus or minus 100 years, which makes the guess of no value in setting public policy.

EARTHQUAKE PREDICTION

Reputable scientists have had major difficulty in answering the question, "when's the next Big One?" and so non-scientists step into the media spotlight. The late seismologist Charles Richter of Caltech was quoted as saying:

> Since my first attachment to seismology, I have had a horror of prediction and of predictors. Journalists and the general public rush to any suggestion of earthquake prediction like hogs toward a full trough.

Charlie did have a way with words.

Nonetheless, the predictors, attracted by the lure of media exposure, are still out there. One who came to see me called herself Sue the Astrologer, an attractive young woman with business cards and a map with mysterious lines drawn on it that told her when and where the next earthquake would strike. Another predictor, probably not influenced by the chance of 15 minutes of fame, was Jerry Hurley, a high-school math teacher in Fortuna, California, and a member of MENSA, the organization for people with high IQs. He would get migraine headaches and a feeling of dread before a predicted earthquake, with the symptoms worsening when he faced in the direction of the impending earthquake. I suspect that Mr. Hurley was a reluctant predictor; he would just as soon someone else had this supposed "talent."

Some of these predictors have PhDs. One of them, Dr. Brian Brady of the US Bureau of Mines, unsuccessfully predicted a major earthquake at Lima, Peru; that prediction is discussed in the section of this book on Lima. One of the more famous predictors was Dr. Eben Browning, a climatologist with a PhD from the University of Texas, who predicted that a disastrous earthquake would strike the small town of New Madrid, Missouri, on December 3, 1990. Browning based his prediction on a supposed 179-year tidal cycle when the moon and planets would line up, and the resulting high tides would cause a huge earthquake exactly 179 years after a series of large earthquakes had struck this region in 1811–1812. His prediction had been made public in a newsletter that he published with his daughter and was also presented in lectures to business groups around the country. Needless to say, he sold lots of newsletters.

Because of the media attention it received, Browning's prediction was evaluated by the National Earthquake Prediction Evaluation Council (NEPEC), a group of earthquake experts selected to advise the USGS and the President of the United States on the validity of an earthquake prediction. NEPEC evaluated Browning's prediction and concluded that it had no scientific merit.

The NEPEC evaluation did not make the story go away, but instead, it enhanced it. Browning got lots of media publicity, including an interview on *Good Morning America*. As the time for the predicted earthquake approached, more than 30 vans with television and radio reporting crews descended on New Madrid. School was let out, and the annual Christmas parade was canceled. Earthquake T-shirts sold well, and the Sandywood Baptist Church in Sikeston, Missouri, announced an Earthquake Survival Revival.

The predicted earthquake date came and went. No earthquake. The radio and TV crews packed up and left, leaving the residents of New Madrid to pick up the pieces of their lives. Browning changed from earthquake expert to earthquake quack, and he became a broken man. He died a few months later.

For a more complete discussion of the Browning prediction, see Spence *et al.* (1993).

THE PROBLEM WITH EARTHQUAKE SWARMS

Most of the time, earthquake professionals are trying to get people to become more aware of the earthquake threat to their region. This is the case in the Pacific Northwest, where I live, which has not had a huge earthquake in more than 300 years. However, when an area is struck by a swarm of small earthquakes, the public becomes very much aware, and paranoia may follow. The earthquake "prediction" at Haicheng in 1975 was part of a swarm of earthquakes that lasted for weeks. The residents of Haicheng became alarmed, and the Chinese government entered the picture. An evacuation recommendation was issued, and the public quickly followed. When the earthquake struck on schedule, many lives were saved. But in hindsight, the Chinese were lucky. There was nothing in the character of the earthquake swarm that clearly identified those earthquakes as foreshocks, or that the date of the earthquake could be determined from the seismic and geodetic data available in advance.

The Abruzzo region of the Apennine Mountains near L'Aquila, Italy, is also subject to earthquake swarms. The L'Aquila region began to experience a swarm of earthquakes in January 2009, and the public became frightened that these might lead to a major earthquake. Part of the concern was that L'Aquila contains many historic buildings that have not been retrofitted against seismic shaking. As the swarm continued, the deputy head of the Department of Civil Protection, Bernardo de Bernardinis, was quoted as reassuring the public that the small earthquakes in the swarm were *releasing* seismic strain energy, thereby reducing the danger of a major earthquake. This view was also presented by Guido Bertolaso, the former head of the Department of Civil Protection. On the other hand, in February a lab technician and amateur seismologist, Giampaolo Giuliani, began warning the public that a major earthquake was coming, based on radon gas anomalies, although this unofficial prediction was not accepted by mainstream seismologists.

A meeting was convened by Bertolaso, including several renowned seismologists, including Enzo Boschi, the highly respected President of the National Institute of Geophysics and Volcanology (INGV). After the meeting, one of the people (not a seismologist) reassured the public, suggesting that people should go home and enjoy a glass of good Italian wine. But on April 8, L'Aquila was struck by an earthquake of moment magnitude 6.3, causing the collapse of many old buildings and 309 deaths. Later, members of the committee, including De Bernardinis, were convicted of manslaughter charges because their reassuring words had misled the public. This was a highly unfortunate outcome and an embarrassment to the Italian judicial system. (Fortunately, this conviction was reversed on appeal for all members of the committee except for De Bernardinis.)

The earthquake swarm caused the public great alarm, the opposite problem in most seismically active regions where the problem is getting the public to take the earthquake threat seriously. But who knows whether an earthquake swarm will lead to a major earthquake, as it did at Haicheng and L'Aquila, or will just peter out on its own? There was nothing special about the swarm earthquakes at Haicheng or L'Aquila that identified the earthquakes as foreshocks. Earthquake swarms that do not lead to damaging earthquakes are common in many parts of the world, including, in addition to the Apennines, Yellowstone National Park, the Hanford nuclear reservation in eastern Washington, the Reno region in Nevada, Mammoth Lakes in the eastern Sierra Nevada of California, the Salton Basin of southeastern California and adjacent northern Mexico, and the rift valleys of Ethiopia in eastern Africa.

PROBABILISTIC FORECASTING

Nonetheless, the question, "When's the next Big One?" is still out there. Indeed, all the major national earthquake programs, including those in the United States, Japan, and China, had earthquake prediction as a major goal when they were established. But many earthquake scientists doubt that the question can ever be answered except for those earthquakes like Haicheng in China (and another one in Greece in 1995) that are preceded by many smaller earthquakes. But, as stated above, most of these swarms of smaller earthquakes are not followed by a large earthquake. We use the word "foreshocks" only in hindsight if they are followed by a larger earthquake. Many of my colleagues argue that the structure of the Earth is too complicated, and we can only get a good look at rocks near the surface, which is not enough information to make a scientifically valid prediction. What to do?

Earthquake scientists decided to take their cue from weather forecasters. The evening TV weather forecaster rarely guarantees that it will or won't rain tomorrow, but instead gives us a *probability*: a 50% chance of rain tomorrow, meaning that there is a 50% chance that it will *not* rain tomorrow. We accept *probabilistic forecasting* of weather with improved satellite imaging, which always shows what has just happened. TV forecasters also show maps of what they expect is likely to happen based on sophisticated computer models created by agencies like the US Weather Service, part of the National Oceanic and Atmospheric Administration. But we accept the limitation that the weather we get is not always the same as the weather they forecast.

By saying *forecast*, we stop using the now-discredited word *predict*, which we have to admit we can't do. Only a few earthquake swarms are followed by large earthquakes like the Haicheng and L'Aquila earthquakes, leading to an apparently successful prediction for the Haicheng earthquake. The Chinese spent a lot of time studying the unusual behavior of animals – horses acting strangely, or rats running out on wires. The USGS even considered the possibility that Old Faithful Geyser might get off its schedule just before an earthquake. None of these worked. So the word *prediction* has been replaced with *forecasting*.

Even though earthquake scientists are not yet able to predict an earthquake, one of the tasks we are asked to do is to advise power companies of the seismic safety of their power plants, including nuclear power generating stations and large hydroelectric dams. We call these major structures *critical facilities*. First we identify all the active faults that might pose a hazard, with greatest attention paid to those faults closest to the plant or dam. This requires us to work out the slip rate on each fault and to estimate the *maximum credible earthquake* (MCE) expected on that fault. We estimate the average return time of an MCE on a given section of fault, recognizing that there is a lot of uncertainty in when the next earthquake will strike our facility. We then ask ourselves, based on this uncertainty, what is the likelihood of an earthquake during the life of our critical facility. We recognize that the next earthquake might be smaller than the MCE and factor that into our estimate. We would like to know more about the record of earthquakes on our fault, either through its recorded history, which is longer than 2000 years for the North Anatolian fault now posing a hazard to Istanbul, Turkey, but more commonly through geological identification of earthquakes in backhoe trenches, the field we call *paleoseismology* (Figures 3.1, 3.2).

Then we do the same for all the other faults that are a hazard to our area of concern. In the Pacific Northwest, for example, this includes the hazard from the Cascadia subduction zone, where the MCE is a magnitude 9, and from crustal faults (like the Seattle fault in downtown Seattle) with a lower MCE but which may be closer to major cities like Portland, Seattle, or Vancouver, or to critical facilities like a major hydroelectric dam.

A *critical facility* must continue to be operational during and after an emergency, or whose failure could threaten many lives. Thus, a critical facility must be seismically strengthened so that the chances of it being heavily damaged are extremely small. A nuclear power plant and a large hydroelectric dam fall into this category.

The Fukushima Dai-ichi nuclear power plant in northeast Japan was a critical facility that failed. The leaking of nuclear materials that followed the March 2011 Tohoku-oki earthquake and tsunami made it necessary to evacuate residents from their homes permanently because of radioactive contamination. After the Tohoku-oki earthquake, Japan saw fit to shut down, at least temporarily, all its nuclear power plants. The plant was designed for a MCE of 8 to 8.4, based on 150 years of seismograph records, but the unthinkable happened. The earthquake they got was a magnitude 9. If they had gone back 1200 years, paleoseismology would have shown that the subduction zone had experienced an earthquake in AD 869 that is now believed to have been as large as magnitude 9. The tsunami that accompanied the 2011 earthquake overtopped their seawall designed to contain the tsunami, and nearly 16,000 people lost their lives (Figure 4.1). In addition, the crippled power plant is still leaking radioactive material, so the crisis continues. The AD 869 earthquake would have been accompanied by a tsunami with similar wave heights (see discussion on Japan in Chapter 9).

To arrive at a probabilistic forecast, we assemble a group of experts and discuss what the level of hazard is likely to be, including experts with divergent opinions. This analysis incorporates the various combinations of faulting and the various estimates of earthquake magnitude and seismic-wave acceleration we can expect on faults affecting our critical facility. Will the subduction zone trigger earthquakes on crustal faults? Will only one segment of a fault break, or will several segments rupture at the same time? Our experts think of all the possibilities and then vote on how much weight to assign to each possibility, using what we call a *logic tree*. We even assume that we haven't found all of the faults that might generate an earthquake affecting our facility, calling an earthquake on an undiscovered fault a *floating earthquake*. This possibility became very real to the citizens of Christchurch, New Zealand's second largest city, in 2010, when it was struck by an earthquake of magnitude 7 on a fault that had not been recognized beforehand.

Earthquake forecasters then take the next step. They recognize that they don't have to limit their probabilistic forecasts to a single critical facility. Their analysis of the critical facility includes an

FIGURE 4.1 Tsunami from the March 11, 2011 Tohoku-oki earthquake of
magnitude 9 in northeastern Japan, overtopping a tsunami-protective
seawall at Heigama estuary, Miyako, which had been designed to withstand
a tsunami from an earthquake no larger than magnitude 8.4. See Chapter 9
on Japan. Source: reproduced with permission from the City of Miyako.

evaluation of the territory surrounding it, which may be presented as
a map and a descriptive table for each potential earthquake fault. The
government could use the same methods and construct a probabilistic
hazard assessment of a region, including not only the critical facility
but the populated areas around it. They have already located many of
the faults in their analysis of the critical facility, and they only need to
extend it to an entire region or country.

The USGS, through a working group, prepares a probabilistic
forecast for the entire United States (see Map 4.1), and, based on an
evaluation by panels of experts, revises the forecast every five years
based on new advances in seismology, paleoseismology (earthquake
geology), earthquake history, and geodesy (the rate at which the crust
is deforming based in large part on the Global Positioning System
or GPS).

For California, a separate group with this objective is called the Working Group for California Earthquake Probabilities (WGCEP; see http://pubs.usgs.gov/of/2013/1165/). This group produces a map, Uniform California Earthquake Rupture Forecast (UCERF). The results of these probabilistic estimates are presented on a colored map, with the higher probabilities colored an angry red or orange and the lower probabilities colored a more soothing green or blue. The probabilities are given for 30 years, the length of a typical mortgage. After its last meeting, the WGCEP concluded that there is a 99% probability of an earthquake of magnitude 6.7 or greater *somewhere* in California in the next 30 years! The likelihood of an even stronger earthquake of magnitude 7.5 or greater during this same time period is 46%, with

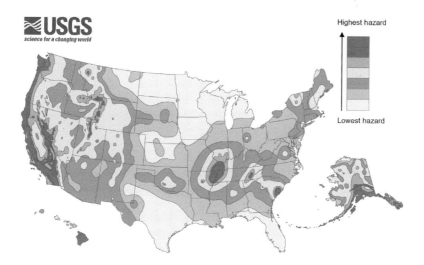

MAP 4.1 Map of the United States showing the 2% probability that the given horizontal acceleration from an earthquake will be exceeded in the next 50 years. The highest probabilities are in the western United States, including the San Andreas fault system (thin dark line in California) and the Cascadia subduction zone. Dark colors in the Midwest mark the New Madrid Seismic Zone, struck by three earthquakes in 1811–1812. The dark color in South Carolina marks the site of a large earthquake in Charleston in 1886. Source: preliminary update to the US Geological Survey Hazard Map, 2014.

the more likely source in southern California. Part of coastal northern California is next to the Cascadia subduction zone, the California section of which is given a 37% chance of an earthquake of magnitude 8 to 9 in the next 30 years (see Chapter 8 on Cascadia). For the east San Francisco Bay Area, the Hayward-Rodgers Creek fault has a 31% chance of generating an earthquake of magnitude 6.7 or greater during the 30-year time interval, about one chance out of three (Figure 4.2). This also means that there are two chances out of three that there *won't* be an earthquake as large as this on the Hayward-Rodgers Creek fault in the next 30 years.

On August 24, 2014, an earthquake of magnitude 6 struck the West Napa fault, a fault that was not included in an earlier estimate of potential rupture, as shown in Figure 4.2. This fault had been evaluated a few years earlier as part of the Earthquake Hazards Reduction Program by John Wesling of the California Department of Conservation and Kathryn Hanson of Geomatrix consulting firm, but their results had not yet been incorporated into the probability map shown in Figure 4.2. Their report did not forecast the earthquake of August 2014.

Probabilities won't alter your vacation plans, but they will affect your insurance rates and should affect your community's building codes and the measures you take personally to strengthen your house and business and your children's schools against an earthquake of this size. Your earthquake insurance rates will go up, or you may be unable to obtain earthquake insurance if you have not bolted your house to its foundation. Forecasts deal not only with magnitude but also with the degree of strong shaking, which for certain regions might be controlled by more than one fault and also by the tendency of soft ground to shake more violently, like a bowl of jello, or of steeper slopes to fail in landslides.

Forecasting was taken up on an international scale through the Global Seismic Hazard Assessment Program (GSHAP), launched in 1992 by the International Lithosphere Program, with the endorsement of the International Council of Scientific Unions. Major contributions

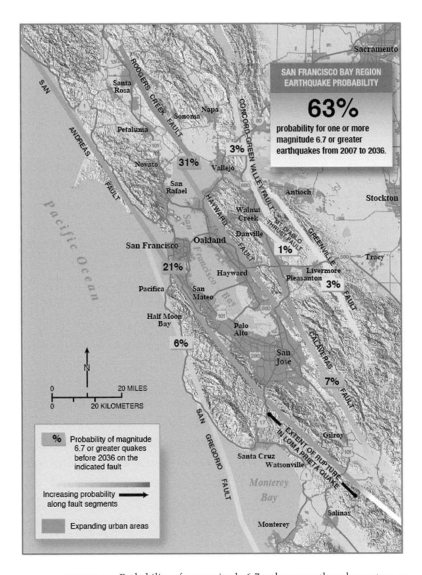

FIGURE 4.2 Probability of a magnitude 6.7 or larger earthquake rupture on active faults in the San Francisco Bay Area in the next 30 years (in this case, 2007–2036). Darker bands indicate higher probability, highest on Hayward-Rodgers Creek fault in the East Bay region. The overall probability for the Bay Area during this time period is 63%, about two chances out of three. This map did not highlight the West Napa fault of magnitude 6 that ruptured near the city of Napa in 2014, causing damage of $1 billion. Source: USGS and Working Group for California Earthquake Probabilities.

were made by scientists from Switzerland, Germany, Italy, Japan, Russia, China, the United States, New Zealand, and other countries. Maps of the world present the 10% probability of exceedance in the next 50 years of maximum ground acceleration, a measure of the strength of ground motion. Again, this also means a 90% probability that these accelerations will *not* be exceeded in the next 50 years.

GSHAP is now being revised in a public–private partnership to create the Global Earthquake Model (GEM), to which I have been a contributor. Before discussing GEM, it is necessary to distinguish between *hazard* and *risk*. *Hazard* involves the probability of earthquake occurrence and shaking at a given location and is based on fault studies such as those described above. GEM, organized by the Organisation for Economic Co-operation and Development (OECD) and headquartered in Pavia, Italy, with support from the World Bank, the insurance industry, and several major international organizations, is creating a digital map of the world's active faults in an internationally consistent format, an evaluation of hazard that can be used for a probabilistic assessment of risk.

The GEM focuses on risk and local involvement (see Chapter 28) The GEM worldwide fault map is probabilistic, like the GSHAP and California maps, but GEM is also based on damage and losses, which should alleviate the underestimation of losses using GSHAP suggested by Max Wyss, director of the World Agency of Planetary Monitoring and Earthquake Risk Reduction in Geneva, Switzerland. The involvement of local communities in GEM means that it is necessary for leaders in cities at risk to have open access to methods developed internationally to strengthen their communities against earthquakes. That is to say, it does no good to have the earthquake problems dictated to leaders of countries at risk by experts in the developed world. It is necessary for local decision makers to take the lead in earthquake mitigation, taking advantage of technology made available to them through GEM. Local leaders need to believe that the earthquake problem is their problem, and they must play a role in its solution. Istanbul, Turkey, is an illustration of this strategy in action (see Chapter 15).

Risk involves a calculation of damage, losses of life, injuries, and the cost of repairs. The financial risk exposure is likely to be greatest in developed countries, with their more valuable structures. In developing countries, the financial risk is likely to be much less, even though the losses of life will be many times greater, a sad commentary on how our society (at least, the earthquake insurance industry) values property more highly than human life. The first version of the GEM was presented in late 2014. The GEM project has been extended for another five years, to 2019.

On July 15, 2013, the uninhabited South Sandwich Islands between South America and Antarctica were struck by an earthquake of magnitude 7.3. It was recorded on seismographs around the world (thanks to its low-frequency seismic waves), but nobody felt it. So, even though it was a big earthquake with hazard, its risk was zero. I am reminded of the tree falling in the forest. If no one hears it, does it make a sound?

OPERATIONAL EARTHQUAKE FORECASTING

Following the L'Aquila earthquake, the Italian government convened an International Commission on Earthquake Forecasting for Civil Protection (ICEF), composed of ten experienced earthquake scientists from nine countries, including, in addition to Italy, China, France, the United Kingdom, Germany, Greece, Japan, Russia, and the United States. The commission was chaired by Professor Tom Jordan of the United States. It was clear to the commission that the standard methods of probabilistic earthquake forecasting, discussed above, did not adequately serve the public in dealing with the L'Aquila earthquake (Jordan, 2013).

The discussion of probabilistic earthquake forecasting discussed above deals only with long-range forecasting, in which time frames of 30–50 years are considered, and average earthquake recurrence is based on long-term slip rate and slip per event, as determined in most cases from paleoseismology. That was not the issue at L'Aquila, where the problem in the earthquake swarm had a time frame of days, not decades or centuries. This leads to *operational earthquake forecasting* (OEF), as described in detail by Jordan *et al.* (2011; www. annalsofgeophysics.eu/index.php/annals/article/view/5350) and summarized by Jordan *et al.* (2014). OEF is time-dependent.

OEF takes into account the large variability in earthquake recurrence intervals, expressed by clustering, illustrated by a sequence of three large earthquakes in the central United States in 1811–1812 at New Madrid, Missouri, a sequence of earthquakes in Christchurch, New Zealand, in 2010–2011, and a surface-rupturing earthquake in northern Japan a month after the magnitude 9 earthquake on March 11, 2011. A long-range probabilistic forecast would not have forecast the earthquake that struck the city of Christchurch in February 2011, only a few months after the first, surface-rupturing earthquake.

The goal of operational earthquake forecasting is to provide communities with probabilistic information about seismic hazards that can be used to make decisions in advance of potentially destructive earthquakes. It operates on the principle that authoritative scientific information about future earthquakes should be made available to potential users, including the public, in a timely manner. The public should become used to obtaining information about earthquake activity, thereby building trust and reducing the gulf between the earthquake experts and the public they serve. Are ongoing earthquakes part of a swarm or are they aftershocks? In addition, information about earthquake probability should be separated from the use of this information in earthquake mitigation and risk analysis. Mitigation includes earthquake drills and strengthening of inhabited buildings against earthquake damage.

Operational earthquake forecasting has been criticized by some earthquake experts, including Wang and Rogers (2014). If OEF increased the predicted probability of an imminent earthquake based on earthquake clustering, it would increase the probability from

0.001% (practically impossible) to 5% (very unlikely) and would have no effect on saving lives. Mitigation would involve evacuation of unsafe buildings, but Jordan *et al.* (2014) argue that a mitigation strategy built only around evacuation of buildings underestimates the potential of OEF in mitigation.

DETERMINISTIC FORECASTING

This is another way of discussing worst-case scenarios, which Max Wyss of Geneva, Switzerland, has suggested would raise awareness among the public in seismogenic areas. It is also a way to forecast superquakes, of magnitude 9 or higher. Professor Chris Goldfinger of Oregon State University and his colleagues, including Professor Ikeda and myself, pointed out several earthquake-prone regions in addition to Japan that have experienced superquakes larger than any normally considered in a probabilistic analysis. Increasing the possible size of the next earthquake would raise the perception of risk to the local population and increase the possibility that local action might be taken simply because local decision makers would need to plan realistically against a much larger earthquake.

The Tokyo Electric Power Company estimated that the maximum size of a subduction-zone earthquake on the subduction zone would be magnitude 8.4, and seawalls were constructed to protect against tsunamis resulting from an earthquake of that size. However, as illustrated in Figure 4.1, the tsunami of 2011 was larger, leading to the deaths of nearly 16,000 people. Several Japanese scientists had warned that the next subduction-zone earthquake could be of magnitude 9, but their warnings were not acted on by the Japanese government. Similarly, the estimates of a magnitude 9 earthquake on the next Cascadia earthquake is a deterministic estimate, although it is estimated that the next earthquake could be smaller, particularly if it affected only the southern part of the subduction zone.

RESILIENCE SURVEY

Another method of considering the next big earthquake is a *resilience survey*, which takes into account not only the damage to buildings and the loss of life from the earthquake, but also the longer-term effects on society. How long does it take for the region to recover economically to its pre-earthquake state? The section in this book on cities in the Pacific Northwest presents the results of resilience surveys in the states of Oregon and Washington, with the stark result that it could take up to a generation for the economy of western Oregon to recover after a major earthquake of magnitude 9 on its offshore subduction zone. If the earthquake were to cut off access to the Oregon coast, utilities could take months to a year to recover. Businesses would have to leave the region to survive, resulting in a loss of jobs and tax revenues. Rebuilding of highways, utility lines, hospitals, police stations, and schools would require a huge financial investment that the states of Oregon and Washington might be reluctant or unable to make. The Congress, currently obsessed with eliminating the national debt, might also be unwilling to step up and help.

So I am back to my interview and question, "When's the next Big One?" I can say, "It could strike tomorrow or a hundred years from now," which is another way of saying "I don't know." The next question could be "Will it be of magnitude 9 or smaller?" Same answer.

Many talented scientists are working on the answer worldwide, and perhaps we will be able to give a more satisfactory answer in the next few decades.

EARTHQUAKE MITIGATION: WHERE'S THE SWEET SPOT?

Much of this book describes earthquake mitigation, commonly involving retrofitting seismically dangerous buildings against the next earthquake. This would be an easier job if we knew when the next earthquake would strike, tomorrow or a century from now. The scientists warning the President of Haiti of an earthquake in the very near future could not tell him when the earthquake would strike, so the retrofitting did not get done, and more than 100,000 people died. The problem was that Haiti, an extremely poor country, could not afford to do the necessary strengthening of its building, choosing instead to gamble that the earthquake would arrive later rather than sooner. The president lost his gamble.

Should seismically dangerous cities be strengthened against earthquakes or is it worth the expense of mitigation? This question was discussed when the USGS was considering the cost of earthquake upgrades in the New Madrid area, in the Mississippi Valley, which had been struck by three damaging earthquakes in 1811–1812. The vulnerability of that region is much greater now, with the presence of major cities like St. Louis, Missouri and Memphis, Tennessee near the footprint of the New Madrid earthquakes. However, unlike California, the region does not appear to be earthquake country, and in the planning meetings that followed, opposition arose to mitigation solutions that would cost billions of dollars. Much of the opposition was led by Professor Seth Stein (2014), a distinguished seismologist at Northwestern University in Evanston, Illinois.

Another example is the cost of retrofitting hospitals in southern California after a destructive earthquake in 1971. The State of California passed a law that required strengthening of hospitals against earthquakes without consideration of how this strengthening would be paid for. It would be counterproductive to force a hospital to do a major seismic upgrade if the result was that the hospital had to close its doors. Accordingly, more than 40 years after the 1971 earthquake, most of the hospitals in California do not meet the higher standard, and the cost of meeting that standard would be at least $50 billion, modified from a cost estimate by the California Healthcare Foundation. Stein was told by one doctor that he could treat a lot of patients for $50 billion! If a hospital is forced to adopt expensive upgrades, who pays?

After the terrorist attacks of 9/11, the US Transportation Security Administration, part of the Department of Homeland Security, with 60,000 employees and a budget of $8 billion, was formed to screen airline passengers, although it is not clear how many terrorists were caught with the increased screening. Also, if the hassle of airport security caused people to drive rather than fly, has the increased number of highway accidents been factored into the costs of mitigation?

Another mitigation strategy is the erection of high seawalls against tsunamis on the Tohoku coast of Japan. Japanese seawalls were designed for a maximum earthquake magnitude of 8.4. However, in 2011 the Tohoku coast was struck by an earthquake of magnitude 9, overtopping the seawalls (Figure 4.1) and resulting in a death toll of nearly 16,000 and the failure of two Fukushima Dai-ichi nuclear reactors. Tsunami deposits show that the last earthquake as large as the 2011 earthquake struck the Tohoku coast in AD 869, more than 1100 years earlier. With a recurrence interval of more than 1100 years, would the higher seawall have been the best investment? (The response to this question by survivors of the 2011 tsunami will probably not be the same as that from mitigation strategists planning for the next earthquake and tsunami.)

The cost of the next earthquake needs to include the cost of mitigation as well as the cost of the earthquake itself. If there is too little mitigation, the cost of the earthquake is higher, and may include smaller earthquakes and smaller tsunamis. If there is too much mitigation (such as in the Transportation Security Administration example), the cost of mitigation exceeds the cost of the hazard (terrorists boarding planes). Where is the "sweet spot"?

5 Population explosion and increased earthquake risk to megacities

INTRODUCTION

Because of plate tectonics, the Earth has been subjected to earthquakes for millions of years. However, in terms of lives lost, the past is not the key to the future. It is only in the last two centuries that Earth's population has been measured in the billions, and the additional people have become concentrated in crowded cities. Most of us are aware of the effects of increased population on the world food supply and on environmental pollution, including greenhouse gases and dirty water. But the unprecedented migration to cities places millions of people at grave risk from earthquakes.

THE POPULATION EXPLOSION

Ten thousand years ago, the world was inhabited by between one million and ten million people (Figure 5.1). Humans were beginning to evolve from primitive hunter-gatherers to farmers and herders, starting in the Middle East and Egypt. Later, the first cities were organized, such as the desert city of Jericho, founded many thousands of years ago near a spring close to the Dead Sea fault. Other cities grew in Mesopotamia, Egypt, Palestine, Greece, India, central China, Mexico, and Peru, and writing and culture developed. The social organization provided by living in cities led to people living in houses rather than caves. The increased availability of shelter and food from nearby farms gave people leisure time for writing and art. The cities flourished.

Hunter-gatherers had to worry about being attacked by wolves and tigers, but the growing cities, because of the problem of many people living close together in unhealthy surroundings, had to cope

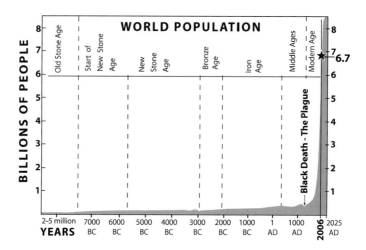

FIGURE 5.1 World population from 2–5 million years ago projected to AD 2025. Most of the population growth has taken place in the past two centuries. Source: Population Reference Bureau (1994).

with disease, with the most famous example being the Black Death plague of medieval Europe. Sanitation, including clean water, was poor to nonexistent, and disease, together with wars, had the effect of keeping the population in check. A modern example of this condition is found among the isolated Kyrgyz nomads of the Wakhan corridor of easternmost Afghanistan: no roads, no government services, no medical care, and a high death rate, particularly among their children. Because of their isolation from the rest of Afghanistan, the Kyrgyz tribes of the Wakhan corridor have not increased in numbers, as documented in a recent article in *National Geographic Magazine* (Finkel and Paley, 2013).

Despite the slow growth, slow because of disease and wars, world population around AD 1 in the early days of the Roman Empire had grown to about 200 million. At about this time, Rome itself became the first city to reach a population of one million. During the Middle Ages, after the fall of Rome, the classical cities of the Mediterranean declined in importance, although cities in India, China, Peru, and Mexico, isolated from the West, continued to thrive.

In Europe, frequent wars took their toll, as did pandemics like the Black Death, which caused world population to decrease from 400 million in AD 1340 to only 350 million a few decades later. A similar pandemic of smallpox and other diseases brought by European explorers in the sixteenth century largely depopulated the Americas. It took several centuries for the world population to reach the level it had achieved in 1340.

In the eighteenth century, scientific advances accompanying the Age of Enlightenment in Europe led to a higher standard of nutrition and medical care. Because more children survived to adulthood, the population began to grow at a more rapid rate. Average life expectancy increased from about 20 years to 60 years. Achievements in medicine, at least in advanced societies, were accompanied by a higher level of sanitation, including clean water. By the middle of the nineteenth century, world population had reached the one billion mark.

The population doubled by 1920 and doubled again by 1975, reaching four billion. By the end of 2011, it had reached seven billion, and it is on track to double again from its 1975 level in the near future. However, this latest increase in population has been at a slower rate, and some developed countries with many elderly people, such as Japan and Russia, are decreasing in population (Figure 5.2). The natural birth rate in those countries is not high enough to exceed the death rate among their old people. This means that, even though the world population continues to increase, it will increase more slowly, particularly in the more advanced countries.

At the beginning of the nineteenth century, only 3% of the world's population lived in cities. By the end of the twentieth century, 47% lived in cities. This had increased to more than 50% by 2010, and current estimates project that 70% of the world's total population will be living in cities by 2070.

However, there is a major difference between advanced and developing societies. Because of the increased percentage of people living in cities, the rural population of developed countries has been

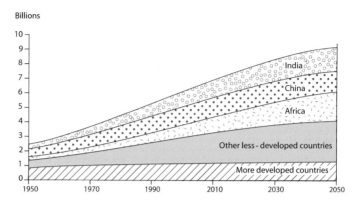

Billions

FIGURE 5.2 Population of various regions in billions in AD 1950 and projected to AD 2050. Greatest growth will be in Africa and India; less growth in China, and other less-developed countries. In contrast, more developed countries show no major growth in this period. Source: United Nations Population Division as estimated in 2006. Based on work by Paul Ehrlich of Stanford University and Population Reference Bureau.

decreasing for more than a half century. The global urban population of developed countries is still increasing, but at a slower rate. The rural population of developing nations is also still increasing, but is close to its peak and should begin to decline in the next few decades.

On the other hand, the urban population of developing nations should continue to increase, although more slowly. This means that more people live in cities in the developing world today than at any time in history, which means that their past losses in these cities from major earthquakes are not a guide to their future losses. The horrific number of deaths in Port-au-Prince, Haiti, in the 2010 earthquake illustrates what is in store for other developing cities close to active faults with a history of major earthquakes. Much of the growth in these cities is taking place in their slums, meaning that, like Port-au-Prince, their population at risk is increasing rapidly. As pointed out by my colleague, Roger Bilham of the University of Colorado, there is a strong possibility that the first natural disaster to take more than a million lives will be an earthquake that scores a direct hit on a major megacity in the developing world.

Changes in population are illustrated by comparing increases in the world's largest cities. In 1975, the world's most populous city was Tokyo, with 26.6 million people. By 2007, Tokyo's population had increased to 35.7 million, but the estimated number by 2025 is only 36.4 million, an indication of Tokyo's slowing rate of increase. The second largest city in 1975 was the New York–Newark metropolitan area, with 15.9 million. By 2007 its population had increased to 19 million, but the number of people living there by 2025 is estimated as 20.6 million, when New York–Newark will drop from the second most populous urban region on Earth to seventh. Other cities in the developed world – London, Paris, Los Angeles, Chicago, Osaka-Kobe, and Seoul – show similar trends.

In contrast, a United Nations study estimates that most of the world's population growth is taking place in the developing world. Consider the population of developing-world cities, some of which I have listed as *earthquake time bombs*. Manila had 5 million people in 1975, 11.1 million in 2007, and is projected to reach 14.8 million by 2025. Tehran had 4.3 million in 1975, 7.9 million in 2007, and is projected to rise to 9.8 million in 2025 based on conservative estimates, or to as large as 12–15 million by other estimates. Karachi, Pakistan, was a fishing town of a few hundred thousand people a century ago. By 2007, it had more than 12 million people, and by 2025, it is expected to have more than 19 million, becoming the tenth largest city on Earth. Some rapidly growing cities in the developing world, like Nairobi in Kenya, did not even exist until the end of the nineteenth century.

The movement of people to large cities in the developing world is the largest-scale human migration in history. People have left their villages where their ancestors had lived for centuries, and they have moved to the big city for a better life, including improved opportunities for jobs. This is the same rationale used by people moving to cities in the developed world, but with one major difference: The migrants to developing-world cities come with their families but with no jobs. The housing they find is in huge slums with very few amenities and buildings with no resistance to earthquakes. These migrants are politically powerless, and commonly even their numbers are not well known by the national government. This leads to a major social problem and a setting for an earthquake disaster.

BUILDINGS CAN BE MADE SAFER, SO WHY DOES THE DEATH TOLL FROM EARTHQUAKES STAY SO HIGH?

Scientists cannot predict when the next earthquake will strike, but engineers have been successful in designing buildings that will protect their inhabitants against earthquakes. Why, then, doesn't the world-wide death toll from earthquakes, about 60,000 per year, decrease in response to our increased ability to make buildings earthquake safe? Why do we still predict that we will some day have an earthquake taking the lives of more than one million people?

It depends on where you live. In northeast Japan, for example, the magnitude 9 Tohoku-oki superquake took many thousands of lives, but these were mainly from the tsunami, because Japanese planners had underpredicted its size. If one removes the fatalities from the tsunami, northeast Japan came out of the earthquake itself fairly well. Japanese towns had adopted building codes that took into consideration the earthquake shaking that they should expect, and retrofitting meant that the losses from building collapse alone were not great. Another superquake, the Maule, Chile, earthquake of magnitude 8.8 on February 27, 2010, took 525 lives and left another 25 missing in the affected area, which contained a population of 12.4 million. The loss of life as compared with the magnitude of the earthquake was relatively small because Chile is a country with modern building codes, many well-engineered buildings, and a public awareness of the hazard from earthquakes based on recent experience, including a magnitude 9.5 earthquake in 1960. Chilean engineers and governmental officials would be the first to point out that they could have done better, but their losses were in the hundreds, not the tens of thousands.

In contrast, the earthquake of magnitude 7 on January 12, 2010, near Port-au-Prince, Haiti, accounted for estimates of lives lost as high as 300,000, although other estimates are much lower, as discussed in greater detail below. Haiti is the poorest country in the Western Hemisphere, with ineffective to nonexistent building codes and a government unable to provide its citizens with basic services. Years after the earthquake, the capital, including the presidential palace, has not been repaired (see Figure 24.2), and many people still live in makeshift camps under lawless conditions.

The Bam, Iran, earthquake of magnitude 6.6 in December 2003 was accompanied by more than 40,000 deaths, even though active faulting at Bam had been known to Iranian geologists for nearly three decades. At about the same time, the California Coast Ranges were struck by the San Simeon earthquake of the same magnitude as the Bam earthquake. But in the San Simeon earthquake, only two people died.

Nicholas Ambraseys of Imperial College London and Roger Bilham of the University of Colorado looked at the question of high death tolls from earthquakes in countries like Turkey and Iran. The civilizations of both countries evolved over thousands of years, and so they would not be classified as developing countries. Both countries have modern universities that have graduated world-class experts in earthquake geology, seismology, and earthquake engineering. Some of these experts have taken jobs in the developed world, causing a brain drain in their country of origin. Why, then, did the August 17, 1999 Izmit, Turkey, earthquake of magnitude 7.6 result in more than 17,000 deaths?

The zone of strong shaking and high damage from the Izmit earthquake extended into the easternmost suburbs of Istanbul, Turkey's largest city, where many of the deaths occurred. People from the rural regions of Anatolia had been moving to Istanbul for decades in search of a better life. To accommodate these new arrivals, apartment buildings had been thrown up, using slipshod methods that were approved by corrupt building inspectors, and these buildings, some relatively new, became death traps. Turkey has building codes, but the salaries of building inspectors were low compared with the cost of the building being inspected, so bribery was rampant.

The high death toll from the earthquake was used by the media to criticize the Turkish government, leading to a claim that the true death toll was much higher than the official number given by government authorities. This claim is controversial, and the problem is discussed in Chapter 15. The people living (and dying) in these poorly constructed buildings, being migrants, had no political clout, and so they themselves had little impact on the discussion about the high death toll. Despite the criticisms by the media, there was no political penalty against those in authority for approving the construction of apartment buildings that were likely to collapse, with great loss of life.

However, the government, guided by Turkish earthquake engineers and scientists, is now dealing with this problem, aided by a major loan from the World Bank, as discussed in the chapter on Istanbul and the final chapter.

Ambraseys and Bilham (2011) described the corruption in the construction industry and among government building inspectors, leading to thousands of deaths. (Ambraseys had first pointed this out in a report to the United Nations in 1976, but his results were not fully published until 2011.) In fact, Ambraseys and Bilham concluded in a paper published in *Nature* that the construction industry, worth about US$7 trillion, is the most corrupt industry on Earth! The poorer the country, the more corrupt its building industry.

Returning to the Haiti earthquake, the losses of life were a reflection of the lack of building standards in the capital city of Port-au-Prince, even in public buildings like the presidential palace (see Figure 24.2), the national cathedral, and the Parliament building. Elsewhere, Ambraseys and Bilham even presented examples of buildings that collapsed *without an earthquake*: in Shanghai on June 27, 2009 and Delhi on November 15, 2010.

A more recent example was the collapse of the eight-story Rana Plaza garment factory in a suburb of Dhaka, Bangladesh (another earthquake time bomb), on April 24, 2013, with the loss of more than 1,100 factory workers who had been trapped beneath the rubble. Construction of the three-year-old building, whose owner was politically well-connected to the ruling Awami League Party, had been permitted for only five stories, and the upper three stories were built illegally. But even if it had only been a five-story building, its poor construction still made it a death trap. There was no earthquake.

Fatalities from shoddy construction of buildings have averaged 18,300 per year since 1980. Other reasons for a high death toll include fire and also building on unstable ground subject to liquefaction and landslides. The highest number of deaths from a single earthquake, 830,000, occurred in 1556 in central China, because many people lived in caves (*yaodongs*) dug into soft, wind-blown silt; these caves collapsed during strong shaking, entombing their inhabitants (see Chapter 12 on China).

Losses from building on unstable ground can be prevented by simply requiring that a geologist and soils engineer evaluate the site prior to construction. This evaluation may condemn the site from being built on at all, thereby greatly reducing its value, a political problem if the landowner has influential friends in the government. Ordinances requiring an evaluation of a site for its stability are enforced in only a few developed countries like Japan and New Zealand, and a few Western states like California and Utah.

How can corruption be quantified? The degree of corruption in different countries has been evaluated since 1995 by a nonprofit organization called Transparency International, headquartered in Berlin (see www.transparency.org/policy_research/surveys_indices/cpi/2012 /indetail4). Transparency International has developed a Corruption Perception Index (CPI) based on polls from institutions knowledgeable about the bribes paid in various countries.

In this book, I use the CPI developed by Transparency International, updated annually, as a measure of corruption in individual countries, although the CPI may measure corruption in fields outside the construction industry, such as drug trafficking. Ambraseys and Bilham observed a correlation between CPI and relative wealth of the country, although they recognized that figures about national income per capita are hard to verify. They recognized that the poorer the country, the worse its CPI, although there are other factors at play.

Some countries are more corrupt than would be predicted just from their relative wealth. These countries include the Philippines, Indonesia, Afghanistan, Haiti, Venezuela, and Iran. Afghanistan is one of the world's most corrupt countries (see Chapter 17 on Kabul). Some of these countries are not rich, but wealth and political power are concentrated in the cities, and their construction industries are riddled with corruption. Scientists cannot tell the authorities when the next large earthquake will strike, so government decision-makers pay no political penalty for not reinforcing their cities against earthquakes. Even when some idea about the return times of earthquakes is known, this earthquake return time is very long in comparison with the terms of office of political leaders with the responsibility for the safety of their citizens. So even when they have some knowledge about the earthquake hazards to citizens who are their responsibility, they gamble that the expected earthquake will not strike during their time in office.

Earthquake recurrence times are long enough that political leaders usually win this gamble. However, in 2010, Rene Preval, the President of Haiti, lost his.

We use the number of deaths as a measure of how destructive an earthquake is, but in the developing world these numbers are suspect. I learned this on an assignment evaluating the destructive Nahrin earthquake in northern Afghanistan, which caused the greatest loss of life of any earthquake on Earth in 2002. The government had one figure, the local authorities had another, and the United Nations had a still different figure. The figure we eventually used, 1200 dead, was based on visiting most of the damaged populated areas and calculating the losses after talking to the chiefs of local villages and towns. But even here, we visited destroyed sites that had comprised extended single-family units in which everyone had died in the earthquake, and so there was no one left to tell us how many people had lived there before the earthquake. We were not just gathering statistics, we were documenting a tragedy.

We visited a village in Balochistan in Pakistan that had been damaged in a recent earthquake. In this case, the locals assumed that we were sent there by the Pakistan government, and so they inflated their estimates of losses in hopes that they would receive a higher level of government support to rebuild, including reduced taxes.

The January 12, 2010, Haiti earthquake, with great loss of life in the capital city, Port-au-Prince, is another example of the uncertainty of loss estimates. As reported by Ambraseys and Bilham, the official death toll was reported as 150,000 on January 24, but by February 10 the number had been inflated to 230,000. Later estimates, commonly part of media events without any additional evidence, were as high as 300,000. A district-by-district analysis could not justify a number higher than 52,000 people killed in collapsed buildings, with an additional 30,000 still buried beneath rubble.

The April 18, 1906 San Francisco earthquake, the most costly in US history in 2015 dollars, had an official death toll of fewer than 500 people. Even at the time, that figure was thought to be too low. Jack London wrote in *Colliers* magazine that "an enumeration of the dead will never be made." A study published in 1989 based on a detailed analysis of public records estimated that at least 3000 people had died in the earthquake (Geschswind, 2001). But at the time of the earthquake, California was in the midst of a building boom involving massive land speculation, and major business interests conspired with the Governor of California and the transcontinental railroads to agree on lower fatalities and to claim that most of the property losses in America's ninth largest city were due to fire, not earthquake (Figure 5.3). Corruption is not limited to the developing world.

FIGURE 5.3 Center of San Francisco, burning because water mains were damaged by the 1906 earthquake. Photos like this one were used by business interests to convince investors and the general public that the San Francisco disaster was caused by a fire rather than an earthquake. Source: image from Preston Hotz collection, included in USGS Professional Paper 993 (1978).

Why these differences? For Haiti, the higher death toll could be used by the government as an argument for a higher level of international relief. For the Izmit, Turkey, earthquake, the high number of deaths became an embarrassment for the government and led to attempts to lower the estimates, followed finally by efforts to fix the problem. Complicating the situation in developing countries is the lack of reliable pre-earthquake census figures, meaning there was no official estimate of how many people had lived in an apartment building before it collapsed.

I have used estimates of fatalities, recognizing that these numbers are probably not accurate in the developing world, but are as good as we are going to get. But estimates of building damage are even less reliable. Was the building completely destroyed or could it be restored to use? Are claims of building losses accurate, recognizing that these claims will be used for compensation by the government?

POPULATION AND LOSSES FROM EARTHQUAKES

Many cities at risk from earthquakes are built close to active faults, in some cases plate-boundary faults. Determining the earthquake risk from such faults could be based on the losses from the previous earthquake on the fault. But the previous earthquake is likely to have struck the same part of the fault hundreds or thousands of years ago, when the population of the city near the fault was small. For example, the 2010 Haiti earthquake was two and a half centuries after the previous earthquake on the plate-boundary Enriquillo fault. The eighteenth-century earthquake took no more than 200 lives because it struck prior to the great Haitian population explosion of the past 100 years.

The Chinese earthquake of Shaanxi Province in 1556, causing the largest death toll in history from an earthquake, struck a fault for which the time of the previous earthquake is unknown. There are other faults nearby that show signs of being as active as the 1556 rupture, but there is no evidence that any of them is about to generate a large earthquake.

In the Caribbean, the Oriente fault, just off the southern coast of eastern Cuba, marks a plate boundary, and its eastern extension in the Dominican Republic, the Septentrional fault, has documented evidence of earthquakes in the past 2000 years. But the recorded history of the Oriente fault is too short, and, because it is entirely offshore, there is no chance to determine its potential for an earthquake. Cuba in general is not considered by its leadership to be earthquake country, so earthquake preparedness is certainly not on President Raúl Castro's to-do list. One of the facilities at risk from an Oriente fault earthquake is the US military prison at Guantánamo Bay. What would be the American response if that prison collapsed during an earthquake, killing many of its inmates accused of terrorism?

Most of the recent increases in population on Earth are taking place in developing countries, particularly rapidly growing cities. Housing in these cities is commonly in slums, and the quality of construction is very low. The growing demand for housing is accompanied by cutting corners, an indication that the construction industry in the developing world is corrupt, and political leaders show very little concern for the safety of the newly arrived migrant population. The combination of rapid population growth of cities, the corrupt building industry, and the lack of concern for the lives of politically unconnected citizens by their leaders means that losses from earthquakes in the developing world will be large.

Before you conclude that what is done in developing-world cities is not your problem, consider the following scenario. A major earthquake strikes the Palestinian West Bank near the Dead Sea fault. Because Israel is a developed country aware of earthquake hazards, with strong building codes, the Israeli population survives the earthquake fairly well. But the developing cities of the West Bank and Gaza Strip are not so fortunate, and thousands of Palestinian Arabs die in the earthquake. The difference in losses between Israel and the West Bank suddenly becomes political throughout the Middle East. All of a sudden, it's the world's problem, meaning that it's also your problem and mine.

PART II Earthquake time bombs

INTRODUCTION

The second part of this book describes some of the earthquake time bombs that threaten large populations around the world. Some are selected because of their cultural history, including the fourteenth- and fifteenth-century earthquakes in central China, the 1755 Lisbon, Portugal, earthquake, and earthquakes in Palestine, including some described in the Bible. Some are discussed because the cities are in the news for other reasons, such as Kabul, Tehran, and Caracas. Caracas is an example of a city in which its leaders take no interest in the earthquake threat the city faces, and, instead, appear to go out of their way to ignore the problem.

Caracas is an end-member city well known for taking no steps against earthquakes. In contrast, Istanbul is a city with a violent earthquake past, most recently the pair of earthquakes in western Turkey in 1999, but Istanbul is taking major steps to strengthen the city against the next earthquake. Those steps are a work in progress, and the corruption that has been so costly after past earthquakes may frustrate the city's efforts to upgrade its buildings and protect its citizens against earthquakes. Time will tell if Istanbul is successful.

On the other hand, the first time bombs discussed in this section are cities that have been taking steps against earthquake damage and have provided leadership elsewhere in the world. These include cities in California, the Pacific Northwest, Japan, New Zealand, and Chile. In all of these places, the paradigm shift has occurred among the scientists and engineers living there, but for various reasons, the general public is not fully engaged in the problem and its solution. This may come as a surprise to people living in each of these cities, especially in California. But in places like Los Angeles, the job is not done, and the millions of people living there are at great risk, in part because of budget cuts and loopholes in existing laws and bending the rules to accommodate the rush to urban development. In Los Angeles, the mayor has taken on this problem with technical assistance from the USGS and committees within the mayor's office and from other specialists, and if the earthquake takes its time striking southern California, the city may get lucky.

Will the next time bomb strike one of the cities discussed here? Possibly, but there is a good chance that it will occur in an unexpected place. Readers of this book may think of cities that are of greatest personal concern to them because they or family members live there, or their businesses depend on an economy that cannot be severely disrupted by a catastrophic earthquake. My hope is that the principles described here for the earthquake time bombs that are included in the book can be used by you to look at your own situation and hold those making (or failing to make) decisions on your behalf fully accountable.

Time bombs where the problem is understood, but the response is still inadequate

6 San Francisco Bay Area

The discussion begins with San Francisco because that region had to confront its earthquake hazard much earlier than did Los Angeles. San Francisco became a boom town starting in 1848, when gold was discovered at Sutter's Mill in the foothills of the Sierra Nevada, followed by the arrival of the trans-continental railroad in 1869. In 1857, before the arrival of the railroad, the central California Coast Ranges were struck by the Fort Tejon earthquake of magnitude 7.8, but this earthquake struck thinly populated regions in the back country and had relatively little impact on California history. However, a few years later, on October 8, 1865, the "Great San Francisco Earthquake" struck the San Francisco peninsula, with its epicenter south of the city, near San Jose. Mark Twain, a journalist who had just been fired from his job at the San Francisco *Morning Call*, wrote about his experience in that earthquake to revive his sagging fortunes. Three years later, on October 21, 1868, another "Great San Francisco Earthquake" struck near Haywards (now Hayward) east of the Bay. At least 30 people out of a total population of 150,000 lost their lives. Unlike the Fort Tejon earthquake, the 1865 and 1868 earthquakes struck heavily populated areas and got the attention of the general public, as well as the concern of people on the East Coast who had been interested in investing in development of the new state of California.

The Chamber of Commerce appointed a committee of wealthy merchants to assure potential East Coast investors that well-constructed buildings were safe. A committee of scientists appointed by the Chamber of Commerce documented damage of about $1.5 million, but under pressure from the business community the report written by scientist George Davidson, chairman of the committee, was never published for fear of scaring off investors. Author Bret Harte wrote in 1866, after the first "Great San Francisco Earthquake," that "there is more danger of the concealment of facts, or the tacit silence of the public press on this topic, than in free and open discussion of the subject." An excellent point, but an open public discussion of the two earthquakes was not to be.

Forty years later, on April 18, 1906, the Bay Area was rocked by an earthquake of magnitude 7.7, accompanied by surface rupture on the plate-boundary San Andreas fault. Damage extended from Humboldt County to the north to San Benito County to the south. The earthquake ruptured water mains so that firefighters in San Francisco were unable to extinguish major conflagrations that enveloped the city. The fires were used by politicians and business interests to argue that the devastation was caused by fire, not the earthquake (see Figure 5.3). Other cities in the United States, including Chicago, Boston, and Baltimore, had also suffered major fires, so the politicians argued that there was nothing special about California that made it unusually susceptible to disaster. The official death toll was fewer than 500 people, but a subsequent investigation decades later showed that the death toll was at least 3000, making the earthquake the most destructive in the history of the United States. The low death toll and the claim that the disaster was a fire were parts of a massive cover-up of the earthquake because California was in the midst of a land speculation involving the Southern Pacific Railroad, with the support and connivance of the Governor of California.

Unlike the nineteenth-century earthquakes, the 1906 earthquake was the subject of an investigation by scientists and engineers sponsored not by the federal or state government but by an endowed private research organization, the Carnegie Institution of Washington. This investigation was led by the most famous geologist of his day, Professor Andrew Lawson of the University of California Berkeley. Lawson included in the State Earthquake Investigation Commission the leading scientists of that era, including Grove Karl Gilbert of the USGS, John C. Branner, later to become president of Stanford University, Harry F. Reid of Johns Hopkins University, who would publish in the Carnegie report his theory of elastic rebound after earthquakes, and Fusakichi Omori of Japan, at the time the world's leading seismologist. Lawson included mapping the San Andreas fault throughout the state in the investigation, even though his views on the sense of slip on the fault were later proven to be wrong. One positive outcome was that earthquake scientists under the leadership of Branner and other members of the commission formed a new society, the Seismological Society of America, the headquarters of which is still in the Bay Area.

However, the earthquake and the investigations that followed did not lead to any meaningful planning against future earthquakes, although the emphasis on fire did cause the upgrading of fire codes. Members of the commission were interviewed by the media, which consistently misinterpreted their remarks to downplay the danger from future earthquakes. As a result, the most disastrous earthquake in the history of California did not lead to any strengthening of building codes or preparation of the public against earthquakes. The business community, fearful of frightening away potential investors, had won the round. It was a missed opportunity.

Frustrated by how their statements to the press had been taken out of context, several of the scientists participating in the Carnegie study, including Gilbert of the USGS, spoke out against how they had been misquoted in support of a conclusion that was the opposite of what they believed. Professor Bailey Willis of Stanford University, probably the most famous earthquake scientist of his day, took his frustration one step further and issued an earthquake prediction in 1925 that, because of his stature, was widely reported in the press. The business interests found their own "expert" who downplayed Willis' prediction (which, indeed, he lacked conclusive evidence to have made), and the 1906 earthquake became yesterday's news.

The 1906 earthquake had other effects on the economy of the United States. As reported by Jones (2015), the earthquake caused a reduction in American GDP of 1.5–1.8%. Earthquake losses were partly covered by British insurance companies, which subsequently raised their rates and began to discriminate against American finance bills. This forced the US economy into recession and led to a financial crisis in 1907, the year after the earthquake. Over the next two decades, San Francisco lost population and subsequently had limited growth compared to Los Angeles.

The next major earthquake to strike the Bay Area, the Loma Prieta earthquake of magnitude 6.9, took place on October 17, 1989, well known because it took place during the World Series, and the earthquake was broadcast on television; 63 people lost their lives. On the plus side, people in the stands did not panic; earthquakes are part of the make-up of Californians, and they behaved responsibly during the strong shaking they experienced. Strong ground motion damaged the Cypress Viaduct, causing it to pancake and trap cars and drivers between the upper and lower decks, adding to the lives lost. In addition, the Loma Prieta earthquake caused major fires in the Marina District near Fisherman's Wharf. Debris from the 1906 earthquake had been dumped in the area to prepare San Francisco for a major exposition in 1915 to show the world that "San Francisco is back." Sadly, the dumped material, called "made land," underwent liquefaction so that underground gas lines broke and caught fire, causing major damage.

In 1990, the year after the Loma Prieta earthquake, the legislature passed the Seismic Hazard Mapping Act (see Chapter 7 on Los Angeles), which required the State Geologist to publish maps showing areas where amplified ground shaking, liquefaction, or earthquake-induced landsliding might be expected. The Act required agents selling real estate to disclose to potential buyers if the property was located within a zone of increased hazard as defined by the Act. However, in a 2004 update, the advisory board created to monitor the Seismic Hazard Mapping Act concluded that the 2001 update of the California Building Code adequately addressed the amplified ground-shaking hazard described by the Act, and therefore it was not necessary for the State Geologist to create any additional maps specifically under the Seismic Hazard Mapping Act. This had the effect of shutting down the program by the California Division of Mines and Geology to map active faults, as discussed below under the Alquist-Priolo Act.

The Loma Prieta earthquake led to a major program involving scientists, engineers, planners, and building owners to strengthen the Bay Area against earthquakes. As described by Professor Mary Lou Zoback of Stanford University, utility and transportation managers invested more than $25 billion in upgrading critical infrastructure, a cost borne by rate payers and taxpayers. More than $30 billion has been devoted by businesses and taxpayers to commercial buildings and residential structures against seismic shaking. Several local communities passed ordinances requiring mandatory retrofits and safety inspections, including financial incentives. San Francisco not only recognized its earthquake problem, it found money to strengthen its buildings, thereby strengthening its resilience against the next major earthquake.

In 2014, the Napa Valley north of San Francisco was struck by an earthquake of magnitude 6, accompanied by surface rupture on the West Napa fault. Although the fault had been described by competent observers a few years before, it did not appear on a hazard map published by the USGS (Figure 4.2). No one was killed, but the damage reached at least $1 billion.

7   Los Angeles

After the 1906 earthquake, the scene shifted to the Los Angeles metropolitan area. Unlike San Francisco, Los Angeles was considered by the general public as relatively free from earthquakes, although earthquakes were felt in the desert east of Los Angeles. Earthquakes had taken place in Los Angeles as early as 1769 during the Portolá exploratory expedition and again in 1800 and 1812, when earthquakes severely damaged the San Juan Capistrano mission south of Los Angeles. However, these earthquakes took place when California was still Alta California, a territory ruled by Spain. These events took place too long ago to be part of any conversation about earthquake hazards, although an earthquake in 1920 damaged the Inglewood district near the soon-to-be-discovered Newport–Inglewood fault. After that earthquake, despite the efforts of John C. Branner of Stanford, the propaganda machine of Los Angeles developers swung into action to reassure investors that Los Angeles had nothing to fear from San Francisco-type earthquakes! But then, an earthquake of magnitude 6.4 hammered the small resort city of Santa Barbara in 1925, taking 12 lives, but again the business interests minimized awareness of the damage and danger. People living in Santa Barbara did not appreciate their losses being downplayed in this way, even by their own Congressional representative.

However, there was now a research institution in southern California interested in earthquakes: the California Institute of Technology (Caltech), including engineers, seismologists, and geologists. Its president, Robert Millikan, a future Nobel Prize-winning physicist, was a major influence on a recently graduated physicist-turned-seismologist named Charles Richter.

On March 10, 1933, the Long Beach earthquake of magnitude 6.4 struck the Newport–Inglewood fault, taking 120 lives and inflicting damage of more than $400 million in today's dollars. The propaganda machine, led by the Los Angeles Chamber of Commerce, dutifully rolled out its usual soothing remarks, but it was now the middle of the Great Depression, and resistance was weak because of the collapse of the real-estate market. In addition, the earthquake severely damaged schools in the region, and outraged parents went directly to Sacramento, the state capital, to demand action. The result was the Field Act and Riley Act, with the Field Act the more important since it was enforced at the state level. Even though it was mainly limited to school buildings, it did what the 1906 earthquake failed to do: It led to strengthening public buildings against earthquakes, including the first effort to eliminate unreinforced-masonry (URM) buildings from the California inventory. In addition, Caltech added more world-class experts in earthquake science, becoming one of the major earthquake research organizations in the world. Because the Caltech campus is in Pasadena, these scientists and engineers took leadership roles in earthquake research in southern California.

In 1947, the US Coast and Geodetic Survey asked California structural engineers for advice about strengthening public buildings against strong shaking, leading to the design of strong-motion seismographs that would measure the response of buildings to earthquakes. This led in 1948 to the formation of the Earthquake Engineering Research Institute (EERI), which provides a link between earthquake seismologists and structural engineers.

In addition, with the onset of the Cold War, the federal government became interested in seismographs that could distinguish between natural earthquakes and underground nuclear explosions by the Soviet Union. In 1950, the Federal Civil Defense Administration took charge of preparation of American cities against nuclear attack. This organization became the Federal Emergency Management Agency (FEMA), which later became the lead agency for federal earthquake response programs in the United States. After the terrorist attacks of September 11, 2001, FEMA became part of the new Department of Homeland Security.

On February 9, 1971, the San Fernando Valley, a suburb of Los Angeles, was struck by the Sylmar earthquake of magnitude 6.7, which was accompanied by surface ruptures that cut across housing subdivisions (Figure 7.1), as well as through public buildings. The fault generating the surface ruptures was not known to be a seismic hazard prior to the earthquake.

While working for Shell Oil Co., I had studied the subsurface geology of similar faults in the nearby Ventura basin, which, since I had become a geology professor in Ohio, got me involved in earthquake hazards research for the first time. After the earthquake, I was hired as a summer intern by a consulting firm, F. Beach Leighton and Associates, to study surface rupture and damage from the Sylmar earthquake. I worked on damage to the Olive View hospital, which

FIGURE 7.1 House in Sylmar in the San Fernando Valley, damaged by surface rupture on the previously unknown San Fernando fault. The bump in the front lawn and damage to the curb are caused by surface rupture. Widespread damage from surface rupture led to passage of the Alquist-Priolo Earthquake Fault Zoning Act, regulating construction on or adjacent to active fault zones.

had been open for under four months when it was destroyed by the earthquake. The hospital was a nonductile reinforced concrete structure that had been thought to be safe from strong shaking, although the earthquake proved otherwise. The raw power of the earthquake made an indelible impression on me, and I decided that earthquakes would be the focus of my future geological investigations. As a result of the Sylmar earthquake, the California legislature passed a law called the Alquist-Priolo (A-P) Special Geologic Studies Act, subsequently renamed the Earthquake Fault Zoning Act, which regulated construction on and near active fault zones, where the most recent surface rupture could be shown to have taken place in the past 11,000 years. This law was signed by Governor Ronald Reagan in 1972. The State Mining and Geology Board was given the responsibility of defining an active fault and determining when A-P regulations would be put into effect. The California Division of Mines and Geology (now the California Geological Survey) was charged with preparing a Fault Evaluation Report for each fault zoned under A-P, a report available to the general public. The Act directs local governments and state agencies affected by the Act to evaluate surface-rupture hazard to prohibit the location of developments and structures for human occupancy across the traces of active faults.

This law was the first anywhere to require evaluation of possible earthquake faults prior to approval of construction projects for development. A-P maps of active fault zones prepared by the State Division of Mines and Geology included an area within 500 to 750 feet of the interpreted fault trace, which developers would be required to evaluate for fault rupture hazard prior to approval of their projects for construction. The developer has to show that the fault is not active, that is, it has not moved during the Holocene, or about the past 11,000 years. An A-P fault also has to be *well-defined*. Unfortunately, the A-P Act was not accompanied by enough state funding to map all the potentially active faults in urban California within a reasonable period of time; decades later, many of the faults have still not been mapped under A-P, as discussed below.

The state also passed a law requiring instrumentation to measure strong ground motion in most of the state, paying for it by a small assessment against building permits. (San Francisco and Los Angeles already had such a program, funded by local government.) Finally, as described above, the state required all jurisdictions to include a Seismic Safety Element in their master plans (General Plans), adding earthquakes to other urban natural hazards in the city planning process. The Seismic Safety Element was later merged into the more general City Safety Element.

How effective has the A-P Act been in regulating development? In the first two decades under the law, more than 500 A-P maps were completed, many on faults of the San Andreas system where the hazard was obvious, but only about 20 have been prepared in the last two decades. One reason appears to be that the entire General Fund allocation to the California Geological Survey was reduced from $9.1 million in 2001 to $2.9 million at present. The larger number was due to a supplementary grant from FEMA after the 1994 Northridge earthquake, to produce liquefaction and landslide maps for a few dozen quadrangles. This grant ended, and the budget reverted back to $2.9 million from the General Fund.

The General Fund allocation does not include enough money to carry out the A-P Act. It was from discretionary money from other projects like the FEMA grant that A-P maps were paid for, since there was no other dedicated funding for this mapping program. This is a classic example of legislative *authorization* vs. *appropriation*. The A-P Act was authorized and signed into law, but it was not accompanied by sufficient funds to place it into full effect.

However, no additional earthquakes with surface rupture have struck major urban areas in southern California since the 1971 Sylmar earthquake, although there have been surface-rupturing earthquakes in the thinly populated desert regions to the east. Even with its limitations, the Act has not been popular among some real-estate agents and developers, who argued that the Act prevents the development of one's property and serves mainly as a bureaucratic roadblock.

I found that the A-P Act had very little impact on the Ventura Basin north of Los Angeles, where I had gained much of my own experience. Even when a landfill near the city of Santa Paula showed evidence for Holocene surface faulting, the Ventura County Regional Water Quality Management Board approved the landfill anyway. Was the A-P Act a paper tiger? Not in this case because the landfill did not contain structures for human occupancy, which would have controlled its regulation under A-P.

It is important to point out what the A-P Act does *not* do. It regulates only surface rupture, and, as pointed out above, no urban earthquake in the Los Angeles metro area, including the 1994 Northridge earthquake, has been accompanied by surface rupture since the Sylmar earthquake. One major reason for this is that many of the earthquakes around Los Angeles occur on reverse faults, where one side moves upward and over the other side. Even some of the faults that have generated strike-slip earthquakes in the past have released strain on reverse faults that are called *blind faults* because they do not generate surface rupture. In addition, if the fault is not well defined at the surface, the A-P Act does not apply.

Furthermore, the Act only applies to buildings slated for human occupancy. Older structures pre-dating the passing of the A-P Act are not covered unless the development plan includes major remodeling, which is covered under the building code. Finally, the building code emphasizes saving lives. The building might be a total loss, but if it does not collapse, and inhabitants can get out of the building alive, the building codes are said to have been successful. I emphasize that these problems are regulated by building codes, not the A-P Act.

The 1994 Northridge earthquake, located in the San Fernando Valley not far from the epicenter of the Sylmar earthquake, occurred on a blind fault, so the A-P Act would not have had any bearing on reduction of losses. The magnitude 6.7 earthquake caused at least 57 deaths, $20 billion in property damage, and an additional $49 billion in economic losses, making it the most costly natural disaster in the United States prior to Hurricane Katrina. This, added to insurance claims after the Loma Prieta earthquake in 1989, had a dramatic effect on earthquake insurance because the damage to insured properties exceeded premiums, and at least one insurance company would have gone out of business after the earthquake if it had not been bought out by another carrier. Because California required companies offering homeowners insurance to also offer earthquake insurance, the cost of homeowners insurance skyrocketed, imperiling the housing market for the entire state.

Although the Northridge earthquake occurred only two decades ago, the situation regarding communications is vastly different. As Jones (2015) has pointed out, the Northridge earthquake struck at the dawn of the Internet era, before the formation of the World Wide Web consortium, before the first web browser, and before cell phones were in general use. Would you be able to call out on your cell phone immediately after a large damaging earthquake of magnitude 7, let alone an earthquake of magnitude 7.8, on the San Andreas fault? Probably not. The circuits might be jammed, and the cell phone towers might be damaged.

Because of the effects on earthquake insurance carriers, the legislature authorized what came to be known as a mini-policy, with higher deductibles, that would cover your residence but not outbuildings. The California Earthquake Authority (CEA), signed into law by Governor Pete Wilson in 1996, allowed insurance companies to spread the risk among private insurance companies by transferring their risk to the CEA. The CEA then purchased $2.5 billion in reinsurance (insurance policies issued to other insurance companies), the largest such purchase in history. The CEA offered the mini-policy with higher premiums and lower coverage. The memories of the 1989 and 1994 earthquakes led initially to a large number of policies being written, but in recent years, without any more damaging earthquakes, the higher premiums and reduced coverage have led many homeowners to cancel their policies. This creates a massive social experiment: What will be the government response to a large urban earthquake if a large percentage of damaged and destroyed homes are uninsured? Would homeowners walk away from their damaged homes and declare bankruptcy? Would the government (state or federal) step in? The large number of foreclosures following the collapse of the housing market during the recent Great Recession of 2008–2010 may be an indication of the response to a large urban earthquake, with many homeowners with destroyed homes but no earthquake insurance.

The previous chapter pointed out the business response to earthquakes in the San Francisco Bay Area, discouraging government response to the hazard until the 1933 Long Beach earthquake, when the protests of parents concerned about unsafe schools could not be ignored in Sacramento. Following the passage of the A-P Act, one might conclude that the battle had been won, and the Act could be applied routinely as additional faults were zoned. However, recent investigative reporting by Rong-Gong Lin II, Rosanna Xia, and Doug Smith in the November 21 and December 30, 2013 and January 9–10, 2014 *Los Angeles Times*, and Adam Nagourney in the November 30, 2013 *New York Times*, indicates that some developers and local elected officials were ready and willing to override protection against active faulting in a proposed project where this can be done legally, arguing that the likelihood of a major local earthquake, with a recurrence interval measured in thousands of years during the lifetime of their development (up to 200 years) is relatively low. The following is based on those newspaper articles, together with reviews by John Parrish, California State Geologist, and an engineering-geology colleague, Eldon Gath of Earth Consultants International.

The $200 million Blvd6200 development (500 apartments and 74,000 square feet of shop and restaurant space) is next to the famous Pantages Theatre in Hollywood, assumed to be south of but close to the Hollywood fault, which had been shown to have undergone surface displacement as recently as 7000–9000 years ago. The development is about 200–500 feet south of the mapped Hollywood fault, depending on which survey map you rely on, and Los Angeles city officials concluded that the development would not present an earthquake hazard, despite protests from people living in the area. A groundwater anomaly south of the mapped Hollywood fault, possibly a barrier related to faulting, was identified on the property, although this feature was not considered important enough by the City of Los Angeles to deny approval of the project or to require paleoseismic trenching.

The developer took advantage of the fact that the California Geological Survey, the successor to the Division of Mines and Geology, had not prepared an A-P map of the Hollywood fault. One reason for this, according to John Parrish, the California State Geologist, was that soon after passage of the A-P Act, the City of Los Angeles had agreed to use its Seismic Safety Element to analyze the Hollywood and other urban faults, leading the California Geological Survey to conclude that its limited budget would be better spent outside Los Angeles. However, based on the remarks of Los Angeles city officials quoted below, it would appear that fault zoning in the city was not being strictly enforced, even though John Parrish recommended that even without an A-P map "it would be prudent for any developer that close to the Hollywood fault to dig a trench to locate it." Work on the Hollywood fault had been done, including boreholes under the direction of Professor James Dolan of the University of Southern California. The Hollywood fault is marked by a steep south-facing slope or scarp that, if it had been investigated by paleoseismic trenching, could have provided evidence for active faulting. Alternatively, trenching might show that this scarp was not a fault but a Pleistocene wave-cut sea cliff.

The California Geological Survey has now belatedly prepared an A-P map for the Hollywood fault, responding to the lack of enforcement of its own regulations by the City of Los Angeles under the Seismic Hazard Mapping Act or the city's Seismic Safety Element. However, this was too late to affect construction of the Blvd6200 project, which was already in progress.

Luke Zamperini of the Los Angeles Department of Building and Safety said that no earthquake fault study for Blvd6200 was required because the property was far enough away from faults shown on a map used by the building department. But the map the City of Los Angeles was relying on was a 2010 map of earthquake fault activity that shows the general location of faults but is less precise at a city-block scale. "The whole city of Los Angeles has faults running through it," said Mitch O'Farrell, a City Council member who represents Hollywood and voted in favor of another project known as Millennium Hollywood. "The hard question is: Do we halt all development in Hollywood? Do we wait for that 11,000-year earthquake? We are going to go down a very slippery slope if we halt all construction for an earthquake fault that hasn't been defined."

However, in October 2013, Los Angeles Councilmen Tom LaBonge and Bernard Parks were urging the City to review the earthquake safety of more than 1000 older concrete buildings, considered seismically dangerous by engineering seismologist Tom Heaton of Caltech. Their view found some support by Mayor Eric Garcetti regarding building permits for the Millennium Hollywood project.

If the Blvd6200 development near the Hollywood fault is representative of urban California, not much had been learned up to 2013, despite the damaging earthquakes California experienced during the rapid growth of its cities. The response to identification of earthquake hazards on a fault close to or within a major city is not that different from that in the developing world, although there is a large community of informed earthquake scientists and engineers in California who work worldwide as well as in California. Like the developing world, earthquake expenditures are focused on response, after the fact, and not on planning ahead for the next one. As Clarence Allen of Caltech said years ago, "every earthquake is a surprise." Tom Heaton of Caltech remarked that it's too bad they don't interview the people who live in the buildings they plan to erect.

Lucile Jones, a USGS seismologist at Caltech and Science Advisor for Risk Reduction, said at the time that, unlike San Francisco, which recently approved a plan to finance reinforcements for buildings with vulnerable first stories, weaker because of extensive open space on the ground floor, Los Angeles has, until recently, failed to press for the kind of building safeguards that could reduce casualties (Jones, 2015). "When you have a big earthquake, you get a lot done – but after that, it's hard to get traction," Jones said. "It's really great that we are doing this without having to kill people first." "In 40 years, we haven't managed to do anything about retrofitting the existing structures," she added. "The conversations about earthquakes have always been the trade-off between seismic safety and financial considerations."

The approval of the Blvd6200 development project, despite the existence of the strongest regulations against surface rupture in the world, may have been directly related to the earlier decision by the California Geological Survey to depend on the City of Los Angeles to take responsibility for the Hollywood fault through its Seismic Safety Element so that it was not necessary for the California Geological Survey to prepare an A-P map of the fault. Later, when it became evident that the City was not taking full responsibility for urban fault evaluations, the California Geological Survey completed an A-P map for the Hollywood fault and released it on January 8, 2014. The new A-P map shows the approximate location of the Hollywood fault with respect to the Blvd6200 project, now under construction, and the proposed Millennium Hollywood skyscraper project, tentatively approved by the City by a 13–0 vote in the summer of 2013. However, the California Geological Survey now estimates the age of most-recent faulting near the Millennium Hollywood project as Pleistocene, greater than 150,000 years.

In light of the recent publicity, the City of Los Angeles has now carried out its own investigations for the Millennium Hollywood project, even though the City did not require detailed investigations prior to preliminary approval without the requirement of paleoseismic trenching. This approval included a yes vote by the politician representing the site to be developed. Influenced by the presence of the new A-P zone on the Hollywood fault, the City required additional work. In addition, Mayor Garcetti called on the City Council to consider several seismic safety initiatives, including evaluation of concrete buildings and "dingbat" structures with soft stories – that is, ground floors that are largely open and weaker than the upper floors. In a news conference, Garcetti said "The Big One could be a lot worse."

The Hollywood fault is not the only one not to have an A-P zone until now. The Santa Monica fault to the west also has no A-P zone, and a number of development projects are proceeding toward approval without the benefit of A-P, although the California Geological Survey is now working on an A-P zone for that fault.

In the meantime, earthquake professionals are continuing to improve our understanding of earthquake hazards in the eight-county Los Angeles metro region, building on their experience in organizing the Great California ShakeOut (see Jones *et al.*, 2008). The annual ShakeOut exercise, now adopted in other parts of the world, is a public relations exercise that involves millions of people in California in their response to a potential earthquake of magnitude 7.8 on the San Andreas fault east of Los Angeles (shown on Figure 7.2, described by Porter *et al.*, 2011). The estimated damage from this earthquake, based on expert opinion among 13 special studies and six expert panels, was subdivided into building damage, non-structural damage, damage to lifelines and infrastructure, and losses from fire. The present version of the ShakeOut estimates that the earthquake on the San Andreas fault would result in 1800 deaths and $213 billion in economic losses. The numbers would be much higher except for aggressive retrofitting that is now underway. They are as high as they are because retrofitting buildings against earthquake damage, including the problem of brittle welds, is still a work in progress; much more needs to be done.

Tom Heaton of Caltech pointed out that wooden buildings are more likely to survive an earthquake the size of the 1971 Sylmar or 1994 Northridge shocks, but reinforced concrete frame (RCF) buildings did not fare well in those earthquakes, and several collapsed, including the Olive View Hospital. The welded connections failed,

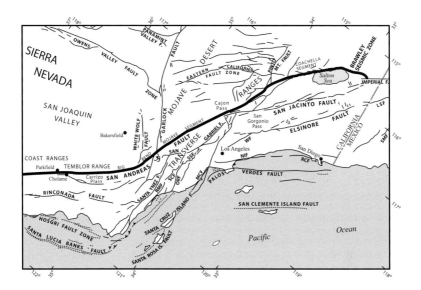

FIGURE 7.2 Active fault map of southern California. The plate-boundary San Andreas fault (heavy line) has the potential for an earthquake of magnitude 7.8, used in the Great California ShakeOut scenario discussed by Lucile Jones of USGS and sponsored by the Southern California Earthquake Center. The last earthquake on the San Andreas fault south of Cajon Pass was more than 300 years ago, indicating a major hazard for Los Angeles, compared with the last earthquake in 1857 on the central San Andreas fault between Cajon Pass and Parkfield. North is to the upper left. Source: modified from Yeats (2012, Fig. 3.7) and fault map by Robert Wallace of USGS.

which was a great surprise to the engineering community. Fortunately, both earthquakes struck early in the morning when those buildings were not occupied, or the death toll might have been much higher. Heaton pointed out that 1400 buildings in Los Angeles still have pre-1994 brittle welds, so the problem has not been fixed. Los Angeles is still a time bomb. This problem was highlighted during Jones' year in the mayor's office, discussed below.

In a public lecture, "Imagine America Without Los Angeles," presented to the American Geophysical Union in San Francisco in December 2013, Lucile Jones presented some of the results of the expert studies leading to the estimates of deaths and economic losses. They include "loss of shelter, loss of schools, loss of jobs and emotional hardship. We are risking the ends of our cities," she pointed out. With the increased importance of the Internet and "just-in-time" delivery of food supplies, the supply networks for Los Angeles could be severely disrupted because the sources of "just-in-time" supplies are on the wrong side of the San Andreas fault from where they are needed. In addition, the concrete water lines that bring water across the San Andreas fault are among the oldest parts of the Los Angeles public-utility network, and many of them are likely to fail in a magnitude 7.8 earthquake. Fiber-optic cables are also likely to fail.

Jones' lecture mirrors issues raised in resilience surveys in the Pacific Northwest, discussed in the following chapter, and in the Global Earthquake Model (GEM), discussed in the final chapter. These studies indicate that the physical losses of buildings and lives are only part of the risk; for California, the entire economy could be affected.

It became clear that the implications of the magnitude 7.8 ShakeOut earthquake could be used for detailed planning by local authorities to strengthen the eight-county Los Angeles region against that earthquake, to graduate from saving buildings and lives to saving the entire economy, also known as *resilience*.

In 2014, 20 years after the Northridge earthquake, everything changed.

The City of Los Angeles signed a Technical Assistance Agreement with the USGS to use the science learned in a century of earthquake research to develop a long-range plan for seismic resilience, a consideration not only of earthquake damage and loss of life, but also economic effects on the city and its population, including recovery costs. The recovery costs from the ShakeOut earthquake of magnitude 7.8 on the southern San Andreas fault had been estimated as $213 billion, or more than $1.5 billion annually, one-quarter of the annualized earthquake costs for the entire United States.

According to the Agreement, Lucile Jones of the USGS, a fourth-generation Californian, would spend all of 2014 in the office of Mayor Eric Garcetti, working on a plan to change the city from an "epicenter of risk" to an "epicenter of seismic preparedness, resilience, and safety." Policy-makers from all divisions of the mayor's office met weekly, using Jones and her earthquake scientific and engineering colleagues from the USGS as technical resources. A Mayor's Technical Task Force was formed that included practicing and academic structural engineers and engineers from the Los Angeles Department of Building and Safety. This task force determined which types of building posed the greatest risk, and drafted proposed ordinances for consideration by the City Council. Technical problems with the water system were addressed by a team from the Los Angeles Department of Water and Power. Telecommunication issues were addressed by the four Internet service providers. Outside experts, building owners, business leaders, and real-estate professionals were asked to address particular issues. The results of these working groups were published by the Mayor's office in a report called Resilience by Design, and released on December 8, 2014 (City of Los Angeles, 2014; Jones, 2015; Rosen, 2015).

The second major change came when Governor Jerry Brown announced his state budget proposal at a news conference in Los Angeles in January 2014, shortly before the twentieth anniversary of the Northridge earthquake. This change affected the California Geological Survey and application of the Alquist-Priolo Act.

The governor announced his determination to increase funding to map California faults as an important public safety measure. His remarks went well beyond responding to surface rupture. He stated, "We'll do whatever it takes. And the people of L.A. should be cautious because earthquakes are just around the corner.... There's a better than 50–50 chance that we'll have a catastrophic earthquake in California that will kill thousands of people and be enormously fiscally devastating." His budget called for $1.49 million in new funding specifically for fault mapping for the next fiscal year, and

$1.33 million in annual funding for future fault mapping, to be paid for by increased building permit fees. At the present time, building permit fees are $10 per $100,000 for residential construction and $21 per $100,000 for other construction. It was proposed to increase permit fees by about 30%, which would also enable the California Geological Survey to use more advanced scientific tools in mapping faults in areas where clues are obscured by trees, buildings, and roads.

Los Angeles joins a small number of earthquake time bombs where decision makers have taken the lead to increase the resilience against earthquakes, joining San Francisco, the Pacific Northwest, and Istanbul. These steps are part of the objectives of the international GEM (see final chapter). However, the question that has yet to be answered is "Who pays?"

This section was reviewed by Lucile Jones in addition to John Parrish and Eldon Gath, acknowledged above. However, any errors in this report are my responsibility

For further reading on the history of California earthquake research, see Carl-Henry Geschwind (1996) (PhD dissertation) and (2001).

8   Seattle, Portland, and Vancouver: Cascadia subduction zone

I had been working on earthquake hazards for seven years when I moved from Ohio to western Oregon in 1977. There was a sense of relief that I could study earthquakes while living in Ohio or Oregon and yet not worry about being in the path of an earthquake myself. Certainly, I knew that Oregon, as part of the northwestern United States, is close to the Cascadia subduction zone, the boundary where the oceanic Juan de Fuca tectonic plate is driving eastward beneath North America. However, in 1977 the subduction zone appeared to be unique in the Pacific Ring of Fire in that seismographs are recording essentially no earthquakes on it at all (see Figure 1.3). In addition, there have been no giant earthquakes on the subduction zone in recorded history, meaning since before the days of the Lewis and Clark expedition in 1804–1806. Some scientists believed that subduction was taking place without building up strain to be relieved by an earthquake. So I could work on your earthquake problem without worrying that it might be my problem as well.

However, my scientific colleagues weren't so sure. The USGS had convened a workshop that I attended in Olympia, the Washington State capital, on a possible earthquake hazard to the Pacific Northwest. Scientists at Caltech, including Tom Heaton, discussed in the preceding chapter, were pointing out that, despite the lack of earthquakes, Cascadia is similar to other subduction zones that have had massive earthquakes, including the Chile subduction zone off southern South America that in 1960 had experienced an earthquake of magnitude 9.5, the largest earthquake ever recorded. That earthquake spawned a deadly tsunami that caused damage and loss of life as far away as Japan (see Figure 8.3). Then, in May 1980, Mt. St. Helens volcano blew off its top, spewing pyroclastic flows and ash, and taking 57 lives. Because of the general view that active volcanoes are an indication of active subduction, the Cascade volcanoes suddenly stopped being simply scenic wonders, and the subduction zone suddenly stopped being a scientific curiosity and became a worry to people living nearby, including local scientists.

Is the Cascadia subduction zone a danger to residents of the Pacific Northwest? In 1987, Don Hull, the State Geologist of Oregon, and I convened a scientific workshop just before the annual meeting of the Oregon Academy of Sciences at Monmouth, including all the scientists working on the Cascadia earthquake problem, to argue about the pros and cons and see if we could come to an agreement. The title of the workshop was: *Is there an earthquake hazard in Oregon or not?*

Don and I expected a lively argument but, amazingly, there was no disagreement! Everyone there agreed that, despite the absence of seismicity on the subduction zone, Cascadia presented a major threat to the Northwest! This major change in opinion is called a *paradigm shift*, similar to the paradigm shift that occurred when Columbus and Magellan proved that the Earth is not flat, but round.

One of the lines of evidence in this detective story, called by some the "smoking gun," was presented at Monmouth by Brian Atwater of the USGS. Atwater had paddled a canoe up the Niawiakum estuary of Willapa Bay in southwest Washington at extremely low tide and found, beneath the salt water marsh grass, roots of trees that were radiocarbon dated as about 1700 years old. He found another zone of tree roots closer to the surface that were 300 years old (Figure 8.1). These are the same species of trees that grow today at slightly higher elevations but cannot grow where the roots are found today because the tree roots are poisoned by salt water. Atwater concluded that the coastal forest had been killed by sudden subsidence, drowning the forest in salt water but preserving the roots of dead trees in the banks of the estuary. His interpretation was supported by microscopic plant

FIGURE 8.1 Marsh deposits in Willapa Bay, Washington, at low tide, with the modern forest preserved at higher elevation. Marshes closer to water contain no trees because their roots were poisoned by salt water when they subsided abruptly in the last earthquake. Two layers of roots radiocarbon dated as ~300 and 1700 years ago are preserved due to abrupt subsidence during subduction-zone earthquakes. Source: photo by Brian Atwater, USGS.

fossils called diatoms, which are very sensitive to water depth, as analyzed by his USGS colleague Eileen Hemphill-Haley.

Still better evidence for sudden subsidence was found farther north along the Copalis River, where a "ghost forest" of dead trees is preserved standing in the marsh grass, poisoned by salt water because of sudden subsidence of the land (Figure 8.2). Radiocarbon samples and analysis of tree rings dated the sudden subsidence at approximately 300 years ago, and Atwater concluded that the subsidence gave the date of the last earthquake, plus or minus a few decades because of the uncertainties in radiocarbon dating. Other marshes in Coos Bay in southern Oregon, Salmon River and Netarts Bay in northern Oregon, and Tofino and Ucluelet in coastal Vancouver

FIGURE 8.2 Marsh grass at Copalis River, north of Willapa Bay. Dead trees (ghost forest) dated at ~300 years old. Modern forest in the background is at a slightly higher elevation so the roots were not poisoned by salt water during subsidence in the AD 1700 earthquake. Source: photo by Brian Atwater, USGS.

Island told the same story that Atwater had found at Willapa Bay and gave the same age of subsidence, about 300 years ago.

Even before Atwater's work on the Washington coast, John Adams, a New Zealander who was then a post-doctoral associate at Cornell University, and his colleagues had been re-occupying geodetic survey benchmarks on highways in the Coast Range and examining tidal records and finding that the Coast Range of Oregon and southern Washington is slowly tilting eastward, toward the Willamette Valley and Puget Sound. They found that at the same time the Willamette Valley and Puget Sound are sinking, the coast is being tilted upward, in the opposite sense as the evidence from the roots, the "ghost forest," and the diatoms, which record subsidence (Figure 1.4, upper diagram). It was concluded that the coast is now slowly uplifting, but at some time in the future it will suddenly drop downward beneath sea level, as it had done 300 years ago (Figure 1.3). The most likely reason for this sudden subsidence is an earthquake on the subduction zone. This interpretation was strengthened by the discovery of sand deposits atop the drowned root zones, sands that Atwater concluded were deposited by a giant tsunami, or seismic sea wave. The root zones covered by sand resembled similar features formed during a subduction-zone earthquake in 1964, of magnitude 9.2 in southern Alaska (Figure 1.4).

At the same time, the assumption at Monmouth that no one had recorded the last earthquake was shattered by oral traditions from

Native Americans living 300 years ago, stories that had been passed down from generation to generation. Here is the account of a Yurok legend by Tskerkr of Espeu, recorded by A. L. Kroeber a century ago:

> Earthquake said, "Well, I shall tear up the earth." Thunder said, "That's why I say we will be companions, because I shall go over the whole world and scare them..." So [Thunder] began to run, and leaped on trees and broke them down. Earthquake stayed still to listen to his running. Then he said to him, "Now you listen: I shall begin to run." He shook the ground. He tore it and broke it in pieces.... All the trees shook, some fell.

Native Americans also knew about tsunamis. This story is a Tolowa legend from northern California recorded by P. L. Goddard in the early twentieth century:

> If the earth shakes east and west, the sea will rise up.... The earth did truly shake from the west and everything on the earth fell down.... [A brother and sister] ran up the hill and the water nearly overtook them.... The water was also coming up the mountain from the east because all the streams were overflowing.... After ten days the young man went down to look about and when he returned, he told his sister that all kinds of creatures large and small were lying on the ground where they had been left by the sea. "Let us go down," his sister said.... But when they came there, there was nothing, even the house was gone. There was nothing but sand. They could not even distinguish the places where they used to live.

On the beach on the west coast of Vancouver Island, the Huu-ay-aht First Nation village of Anacis suddenly disappeared 300 years ago, a casualty of the tsunami accompanying the last earthquake. Although the village was rebuilt after the earthquake, the tribe has now decided to move their community to higher ground, beyond the range of tsunami waves.

TAKING THE MESSAGE TO THE PUBLIC

By the end of the meeting at Monmouth, the scientists recognized that a paradigm shift had taken place, but now they faced a larger and more important paradigm shift. They had to convince the citizens of the Pacific Northwest and adjacent British Columbia that the entire region is under threat from a massive earthquake, and that something needs to be done about it. Linda Monroe, a journalist from the *Oregonian* newspaper, started it off with her coverage of the Monmouth workshop and its paradigm shift, and additional coverage was published by the *Oregonian* in the next few years. I included a description of the Cascadia subduction zone earthquake and tsunami in my textbook *Living with Earthquakes in the Pacific Northwest* (Oregon State University Press) that is used in colleges in the Northwest and is now available online for free. Others prepared videos of the subduction zone hazard that were shown on television. But the challenge was and is huge. People live in the Northwest because of its natural beauty, and endure the long winter rainy days as part of life. But earthquakes? As my next-door neighbor said to me, "Bob, you gotta be kidding me." (He got the paradigm shift, though, and moved to Florida.)

FIRST STEPS: UPGRADE BUILDING CODES

The most important immediate steps were to upgrade building codes. Up until 1974, Oregon building codes did not consider damage from earthquake shaking at all. Oregon was considered as earthquake-safe as Kansas. In 1974, before our meeting in Monmouth, the building codes were strengthened to take into account resistance to some seismic shaking, a response to earthquakes within the subducting Juan de Fuca plate in 1949 and 1965. But it was not until 1993, after the scientific paradigm shift, that Oregon structural engineers were successful in upgrading Oregon's building codes to take into account shaking from a subduction zone earthquake. What this means is that buildings constructed prior to 1974 are not protected against seismic shaking at all, and it was not until 1993 that protection was upgraded to the extent that people, including school children, could count on surviving in a building struck by a subduction zone earthquake.

However, there are two problems with this approach. First, as stated above, the upgrading of building codes is not retroactive to include all structures built before the new standards were implemented. The building codes apply only to new construction or major remodeling of older buildings. If the building where you live or work was built or remodeled earlier than those dates, tough luck. You are not protected. Second, the goal of the new building codes is to get you out of the building, alive and uninjured. But the building itself could be a total loss.

MORE EVIDENCE, BUT WE STILL CAN'T PREDICT THE NEXT BIG ONE

In the meantime, new evidence supporting the idea of a Cascadia subduction zone earthquake continues to be found. Japan, for example, is a country subjected to frequent earthquakes accompanied by tsunamis (a Japanese word meaning "harbor wave"). Professor Kenji Satake, now at the University of Tokyo, discovered an "orphan tsunami" that struck Japan in AD 1700, but was not accompanied by an earthquake. This was described in a book by Atwater *et al.* (2005). Satake examined records written in classical Japanese from coastal settlements up and down the Pacific coast of Japan and realized that the tsunami happened within the same time interval as the last Cascadia earthquake, based on radiocarbon dating (Figure 8.3). By a process of eliminating other subduction zones around the Pacific Ring of Fire (see Figure 1.1), Satake concluded that the Japanese "orphan tsunami" was caused by a Cascadia earthquake. Because the speed of a trans-Pacific tsunami wave can be calculated (about the speed of a commercial jetliner), Satake backtracked the tsunami across the Pacific to the Pacific Northwest and determined that the most recent Cascadia subduction zone earthquake struck on January 26, 1700, about 9 o'clock in the evening! It had a magnitude of 9, based on the size of the tsunami in Japan. The timing was consistent with the time of burial of forests described by Atwater and his American and Canadian colleagues in the coastal marshes. Because it affected

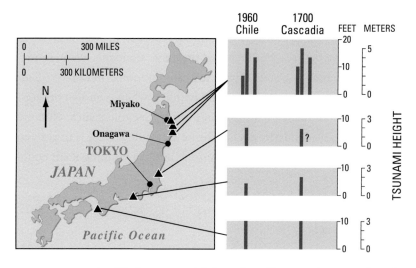

FIGURE 8.3 "Orphan tsunamis" damaging villages on the east and south coasts of the island of Honshu, Japan, from the 1700 Cascadia earthquake of magnitude 9 and the 1960 Chile earthquake of magnitude 9.5. Maximum tsunami wave heights are shown on the right. Japanese historical records allowed the Cascadia earthquake to be dated January 26, 1700, consistent with Native American oral histories. Source: based on research by Kenji Satake of the University of Tokyo, and Atwater *et al.* (2005). Image is from USGS Circular 1187 by Atwater *et al.* (1999).

Japanese keeping records and precise dates, and Native Americans also recorded the earthquake in their oral traditions, the AD 1700 earthquake is now a historical event. The date and time are entirely consistent with Native American stories that the earthquake struck during a winter's evening.

Oceanographic surveys by Oregon State University in the 1960s, under the direction of Professor LaVerne Kulm, included cores from the deep-sea floor of the oceanic Juan de Fuca and Gorda tectonic plates off the Northwest coast. John Adams, now of the Geological Survey of Canada, had attended the Monmouth meeting and followed it up by visiting the Oregon State University core lab to examine the cores taken by Kulm's graduate students two decades earlier. Adams found that these cores include layers of sand called *turbidites* that

he concluded were the result of being shaken loose during a subduction zone earthquake (Figure 8.4). Professor Chris Goldfinger of
Oregon State University followed Adams' lead and took a large
number of additional cores from the deep-ocean floor and dated them
by radiocarbon. These paleoseismological dates are consistent with
the dates from West Coast marshes and provide a history of earthquakes for the last 10,000 years, the post-glacial geological time we
call the Holocene. This is the longest subduction zone earthquake
record on Earth.

Goldfinger found that the entire subduction zone, 684 miles
(1100 km) long, was ruptured by 19 earthquakes in the past 10,000
years, including the AD 1700 earthquake. Several additional

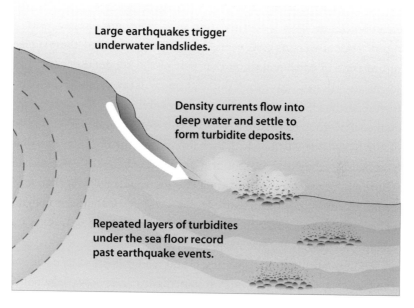

FIGURE 8.4 Strong shaking of the continental margin by earthquakes off
the Pacific Northwest triggers underwater landslides that flow downslope
and are deposited on the deep-ocean floor as sand deposits (turbidites).
The slow rain of fine-grained sediments above and beneath these deposits
allows them to be radiocarbon-dated. Source: based on research by Chris
Goldfinger and John Adams. Figure by Kathleen Cantner, American
Geosciences Institute; used with permission.

earthquakes ruptured only the southern part of the subduction zone in southern Oregon and northern California. Goldfinger's work confirmed that the AD 1700 earthquake was the most recent of a long series of earthquakes. His work showed that ongoing subduction is not a smooth process as was thought by many before our Monmouth meeting, but takes place in sudden jerks, accompanied by large earthquakes.

Problem solved? Not yet.

The problem is that these earthquakes vary in size, with at least two of them larger than the AD 1700 earthquake, based on evidence from the turbidite cores. In addition, the repeat time between earthquakes varies from as large as 1200 years in the northern part of the subduction zone to less than 250 years in the south. This means that scientists can say that *without a doubt*, there will be more great earthquakes at Cascadia, but they can't say when the next one will strike – tomorrow or a century from now. Southern Oregon, with a return time averaging 250 years, has already exceeded the average return time of earthquakes on that part of the subduction zone, but scientists still can't say when the next one will take place. Scientists, including the USGS, have established probabilities of the next earthquake, analogous to weather probabilities (for example, 50% chance of rain in the five-day forecast). Probabilities for the next 50 years that have been established by Goldfinger are a 10% chance of a magnitude 9 earthquake affecting the entire subduction zone, or a 37% chance of a smaller earthquake of magnitude 8–8.4, affecting southern Oregon and northern California. One chance out of three in the next 50 years should get everybody's attention.

These are high probabilities! They won't affect your vacation plans, but they will affect your insurance rates and *should* lead political leaders in the Northwest and British Columbia to retrofit unsafe buildings. One committee in Oregon has recommended that $100 million per year be set aside for this purpose. This would increase the chance that Oregon would be prepared for the inevitable subduction zone earthquake when it strikes.

But the residents of coastal southern Oregon and their political leaders do not appear to be too worried, and they have taken very few steps against their expected earthquake, choosing instead to gamble that it won't happen during their lifetimes. After all, a 37% chance of an earthquake in the next 50 years also means a 63% chance of an earthquake *not* striking the southern part of the subduction zone during that period. Most southern Oregonians, like the President of Haiti prior to 2010, have decided to gamble.

The paradigm shift for the general public has not yet taken place. The people who live in Cascadia are aware of the earthquake problem, but not enough to take decisive action, assuming that it won't happen during their lifetimes.

In contrast, decision makers in Istanbul, Turkey, faced with similar probabilities, are taking major steps to safeguard that city against its next earthquake, as described in Chapter 15. Recent reports on the impact of a great earthquake on the Northwest have led to calls for major retrofitting against earthquakes, but major funding has not yet been allocated. This problem is discussed in the next section.

CONVINCING THE PUBLIC TO TAKE ACTION

This brings us back to the recognition of the earthquake hazard that needs to happen among the people living in the Pacific Northwest. A few years ago, I was at Nye Beach on the central Oregon coast, where I witnessed the local response to a tsunami warning issued by the National Oceanic and Atmospheric Administration (NOAA). As instructed by the city and county, all-terrain vehicles went to the beach and ordered everyone to evacuate. Most did, but as they grumbled about being ordered off the beach, other people arrived with their cameras to take photos of the expected tsunami, which, if it had arrived, could have drowned them all. After the warning, which fortunately was a false alarm, not followed by a tsunami, a Portland television station rated the responses of Oregon coastal communities to the warning and gave several of them a failing grade. This rating enraged the mayor of one of the coastal cities on the central Oregon coast, who was concerned about the effects of the warning on the tourist industry.

I was a consultant for a group of residents of Woodinville, Washington, north of Seattle, protesting the construction of a regional wastewater treatment plant near their homes. According to the International Building Code, such a plant is called a *critical facility*, meaning that special studies and precautions need to be taken because of the potential loss of sewage capability to a large part of metropolitan Seattle in the event of a major earthquake, a loss lasting at least several months. The site lies across one of the largest crustal faults in western Washington, the Southern Whidbey Island fault, which was trenched by the USGS and found to be active because of evidence for liquefaction and possible Holocene faulting.

Despite this evidence, the Seattle Metro executive rammed through approval of the sewage plant, even taking steps on one occasion to prevent specialists from the USGS and Washington State Division of Geology and Earth Resources from viewing trenches excavated by a consultant hired by Seattle Metro. The point here is not the misguided actions of a political leader who no longer lives in Seattle, it is that there was no outrage on the part of the general public, even though the wastewater treatment project was completed way over budget. The Seattle *Times* and local television stations were of little help in presenting an unbiased account of the controversy. Only the Woodinville *Weekly*, a low-budget newspaper that reached only local citizens, took a stand against the project. But how many people read that warning?

On a slightly more encouraging note, the Portland Public Schools, the largest school district in Oregon, placed a $482 million bond issue before the public to, among other things, upgrade old school buildings to strengthen them against earthquakes. An earlier bond issue had failed, but this one passed. The bond issue includes funds for replacement of three of the seismically most dangerous high schools. In addition, there are funds for seismic improvements and roof replacements with seismic benefits for other schools in Portland built before the adoption of building codes that require that buildings be strengthened against earthquakes. It is a good beginning, but they have a long way to go, and time is not on their side.

Seattle schools have had a somewhat better outcome because they had been damaged by non-subduction-zone earthquakes in

1949 and 1965. Following those earthquakes, the school district embarked on a modest program of seismic upgrades, including the involvement of volunteers, with the result that in the 2001 Nisqually earthquake, no school buildings collapsed, and there was no loss of life. Seattle had been lucky in that the earthquakes in 1949, 1965, and 2001 did not occur when the schools were full of students, and the loss of life in 1949 and 1965 from building collapse and falling bricks was minimal. However, the Nisqually earthquake had a magnitude of only 6.8, more than 100 times weaker than the expected subduction zone earthquake of magnitude 9.

In the run-up to the successful 2012 Portland bond issue, an architectural firm proposed an ingenious, low-cost solution: super-strong cages to be placed in the hallways of schools that students could run into if there is an earthquake. This solution is still experimental but worth pursuing if it could save children's lives, but the fact remains that after the earthquake, the school buildings could not be used as emergency shelters, and the only thing left standing would be the cages. The school buildings themselves would have to be rebuilt.

The point to be made here is that the paradigm shift among the citizens of the Northwest is still taking place. Most people have heard that the Northwest is earthquake country, and a few have purchased earthquake insurance. But translating this vague recognition that the Northwest will sustain an earthquake "some day" into positive steps to be sure that the economy survives the earthquake and surmounts its difficulties is unlikely, based on current public attitudes.

The Oregon legislature, led by Senator Peter Courtney, has taken some modest steps, including a "quick look" survey of public buildings, revealing that many are in danger of collapse in an earthquake. The 2011 legislature authorized a resilience study, published in February 2013, to estimate the cost to Oregon of the next subduction zone earthquake. The short answer: The cost will be staggering and may cause a sharp decline of the entire economy of the Pacific Northwest for years, in large part because it would take a long time to get lifelines up and running, as described below. To avoid this catastrophic outcome, the paradigm shift among the general public and the public demand for action must take place *now*, before the earthquake, not at some undefined time in the future.

The problem can be visualized by comparing the effects of a magnitude 9 earthquake in Cascadia and Superstorm Sandy on the Middle Atlantic coast in October 2012. The economic losses from Sandy, estimated a year after the storm as $65 billion, may be larger than those from the Cascadia earthquake because the value of the built environment, including the New York City–Newark megacity, one of the largest in the world, is much higher than it is in the Northwest, even including the cities of Vancouver, Victoria, Seattle, and Portland. But the losses of life could be a hundred times worse. The loss of life in the United States from Sandy, not including losses in the Caribbean, was estimated as 182. The expected loss of life in the Cascadia earthquake and tsunami will be in the thousands, possibly as high as 10,000. One reason is that the US Weather Service was able to predict that the Sandy storm track would turn inland at New York and New Jersey, and people were able to take preliminary precautions, including evacuation. Unless we scientists suddenly get smart about earthquake prediction, the Cascadia earthquake would strike *without warning*. Cascadia would also be accompanied by strong ground shaking lasting several minutes, which was not an issue with Sandy. But the main loss of life would probably be from the accompanying tsunami.

A comparison with the Tohoku-oki earthquake of magnitude 9 in northeast Japan on March 11, 2011, is instructive. Japan's long history includes many earthquakes that have taken tens of thousands of lives, including the 1923 Tokyo earthquake that claimed more than 140,000 lives. Because of its history and its culture, Japan is the best-prepared country on Earth against earthquakes. The losses from the Tohoku-oki earthquake were nearly 16,000, but these were mostly from the tsunami. Losses from strong shaking and building collapse were much lower than they would have been in the Pacific Northwest because most Japanese buildings had already been strengthened against seismic shaking.

The magnitude 9 Tohoku-oki superquake and tsunami that devastated northeastern Japan in 2011 was the same size as the AD 1700 earthquake and tsunami at Cascadia. Coastal Japanese communities subsided during the earthquake, permanently flooding streets near the sea, similar to the subsidence measured by Brian Atwater along the Washington coast, drowning the coastal forest. The Pacific Northwest must plan for a subduction zone earthquake the size of Tohoku-oki.

THE NEXT CASCADIA EARTHQUAKE

Several groups, including Oregon Emergency Management, have attempted to visualize the next Cascadia earthquake through statewide training exercises. In 2011, Jerry Thompson and Simon Winchester presented a graphic and chilling account of this earthquake in a book Thompson published in 2011, called *Cascadia's Fault*. But the Oregon legislature, led by Representative Debbie Boone, elected from a district on the Oregon coast, and Senator Peter Courtney of Salem, wanted something more. What would the losses be, not only in lives or damaged buildings, but in the length of time essential services would be down, and would this cause businesses to flee the state in order to survive? The answers to these questions are so catastrophic that they are almost unimaginable. They include a decline in the economy of the state lasting up to a generation. Although limited to Oregon, the implications apply fully to Washington, which has completed its own resilience survey, northern California, and coastal British Columbia. The resilience survey commissioned by the Oregon legislature illustrated the cost to Oregon of doing nothing or of taking only modest steps.

The resilience survey was directed by the Oregon Seismic Safety Policy Advisory Commission (OSSPAC), which created eight working groups of citizens from government, universities, the private sector, and the general public to examine the impacts on business, the energy sector, transportation and lifelines, water and wastewater pipelines, communications, and critical facilities, with a special focus on the coast. Each group was asked to assess the impact on its respective sector, including the time required to restore function to the way it had been prior to the earthquake. What changes in practices and policies would be necessary for Oregon to reach a desired level of resilience in the next 50 years?

An earthquake of magnitude 9 would cause violent shaking for three to five minutes along the entire Northwest coast from northern California to central Vancouver Island, causing severe damage to unreinforced masonry (URM) buildings and wood-frame houses not bolted to their foundations. The strong shaking would be accompanied by landsliding and liquefaction, rupturing underground utilities, including water and sewer lines. Gas escaping from underground gas lines broken by liquefaction would catch fire, as they did in the Mission District of San Francisco after the Loma Prieta earthquake of 1989, but on a much larger scale.

The earthquake would be followed about 20 minutes later by a tsunami that would be similar to the Tohoku-oki tsunami that caused nearly 16,000 deaths in Japan. The tsunami would overwhelm parts of low-lying towns like Tillamook, Astoria, Seaside, Cannon Beach, Coos Bay, and Newport, ironically including the national tsunami research center at the Hatfield Marine Science Center at Newport, all of which are in the expected tsunami run-up zone. The Oregon resilience survey concluded that the zone of tsunami inundation would include more than 10,000 housing units, with a resident population greater than 22,000. Also inundated would be nearly 1900 businesses employing nearly 15,000 people. The coast would subside permanently five to ten feet, as it did after the last great earthquake in AD 1700 (and as the Tohoku coast of Japan did in 2011). The estimated losses in Oregon alone would be $32 billion.

Washington losses estimated by the Federal Emergency Management Agency (FEMA) would be $49 billion, larger than Oregon because of the higher value of property at risk in the Puget Sound region. Then add northern California and southwest British Columbia, and the economic losses could bankrupt the insurance industry and severely damage the economies of all the affected states and province, as well as the national economies of t he United States and Canada.

The Oregon resilience survey showed that of the 2567 highway bridges in Oregon, 982 were built without seismic considerations at all, and only 409 were designed specifically with Cascadia subduction zone earthquakes in mind. The vulnerability of Northwest bridges was brought home to the public on May 24, 2013, when an Interstate 5 bridge near Mount Vernon, in western Washington, collapsed into the Skagit River due to a truck colliding with the bridge superstructure. Since that disaster, many bridges in the Pacific Northwest have been declared as "obsolete," indicating that examination of bridges should include their resistance to earthquake shaking.

The resilience survey concluded that many bridges on the coast would collapse in an earthquake, isolating coastal communities from one another and from rescue operations from inland cities not damaged by the earthquake (Figure 8.5). Despite the resilience survey, the 2015 Oregon legislature failed to pass a bill to strengthen Oregon's bridges against a subduction-zone earthquake. In the debates by legislators, the hazard from obsolete bridges was not even discussed. Submarine cables would be cut, isolating the Northwest from Alaska and other parts of the Pacific Ring of Fire. The earthquake would instantly damage or destroy electric power lines, natural gas lines, water and sewer systems, hospitals, police stations, and school buildings; 24,000 buildings would be completely destroyed, and another 85,000 would be damaged so extensively that they would take months to years to repair. In summary, the earthquake, even a smaller one affecting only that part of the subduction zone in southern Oregon and northern California, would be a major catastrophe.

According to the Oregon resilience survey, it would take one to three years to restore drinking water and sewer service to the coast and one month to one year to restore water and sewer service to the Willamette Valley and Portland. Restoring police and fire stations in Willamette Valley would take four months, and getting the

FIGURE 8.5 Computer-generated topographic map of the Cascadia subduction zone, with seawater removed. Figure 1.3 gives place names. Subduction zone, at base of the continent offshore, ends at the Mendocino fault at the bottom edge of the image. Onshore to the east (right), the Coast Range separates the offshore and Willamette Valley, Puget Lowland, and Georgia Strait. Roads crossing the Coast Range are likely to be inoperable after a Cascadia earthquake, isolating coastal communities from heavily populated lowland to the east. Gorda sea-floor spreading ridge and Blanco fracture zone show up clearly, Juan de Fuca ridge less so. Black dots: Cascade volcanoes. Source: image by Ralph Haugerud, 2004, USGS Investigation Series I-2689, in collaboration with Geological Survey of Canada, modified by Brian Sherrod, USGS.

top-priority high͟   ͟t partial function would take six
to twelve months͟   ͟cricity and natural gas to the coast
would take three tơ   ͟ths, and restoration of health-care facilities to operational status would take 18 months in the Willamette
Valley and three years on the coast. Comparable estimates of recovery
time were made independently in the resilience survey for the State
of Washington. Both of these surveys are listed in the references at
the end of the book. These resilience surveys were combined by
the Cascadia Region Earthquake Workgroup (CREW) in a single publication, *Cascadia Subduction Zone Earthquakes: A Magnitude 9.0
Earthquake Scenario* (2013), which was independently published by
the geological surveys of Washington, Oregon, and British Columbia.
The CREW report illustrated its findings using the response of Chile
to a superquake of magnitude 8.8 in 2010, discussed below, important
to the Northwest because Chile is a developed country that has
experienced subduction zone earthquakes. This report is sobering,
but should be read by all living in the Northwest to accelerate the
paradigm shift for the general public.

Regarding highways, the main highway between the Willamette
Valley and Newport is now being retrofitted, but failure to recognize
landslides along the route has led to delays in completion of several
years and an increase in the construction costs so large that the
Oregon Department of Transportation considered abandoning the
project after investing many millions of dollars in it.

The length of time to restore services means that those businesses that had not already failed due to building damage, tsunami
flooding, and inability to transport goods from the Willamette Valley
across the Coast Range to the coast could not afford to stay and would
start to leave Oregon and Washington after about a month. As businesses left, people would also leave to follow the jobs to locations not
affected by the earthquake. The result would be a loss in population
and a declining economy that might take a generation to recover.
This, then, is the true cost of not strengthening lifelines and public
structures against the inevitable earthquake.

Smaller-scale examples of where this has already happened include two major storms, Hurricane Katrina in 2005 and Superstorm Sandy in 2012. Katrina and Sandy differ from the Cascadia earthquake in that warnings were issued by the US Weather Service, but there was no effective warning of some of the most catastrophic effects, including the failure of levees on the Mississippi River near New Orleans, the tsunami-like storm surge against coastal towns in Mississippi, and, for Sandy, the storm surge flooding neighborhoods in coastal New Jersey and Staten Island and Long Island in New York State. The death toll from Katrina was 1464 people; 70,000 jobs were lost; and the damage was $128 billion. Evacuation to other parts of the country caused the population of New Orleans to shrink from 445,000 to 312,000 two and a half years later. Removal of floodwaters and repair and strengthening of the levees were (and still are) long term and costly.

Hurricane Sandy was also a catastrophe that will take many months if not years of recovery. A year after the storm, many homes were still in ruins, and homeowners are in litigation with their insurance companies and in arguments with local government over how much help government should provide. Should the coast be rebuilt just as it was, or should a broader section be left undeveloped in the event of another Sandy, as suggested by Professor Orrin Pilkey of Duke University, an expert in the hazards of development along coastlines? The coastal boardwalks of New Jersey cities are the sites of many small businesses, as well as large casinos, indicating major damage to the tourist industry. Will these businesses survive?

The Cascadia earthquake differs from Katrina and Sandy in one more way: The affected area would be hundreds of times larger.

In addition to the subduction zone, the communities of western Oregon and Washington face the hazard of crustal faults (Figures 8.6, 8.7). The largest of these is the Seattle fault, which extends east from Bainbridge Island across Puget Sound through downtown Seattle and the most expensive real estate in the Northwest. This fault sustained a surface rupture, well exposed on Bainbridge Island (Figure 8.8), around 1100 years ago from an earthquake estimated at magnitude

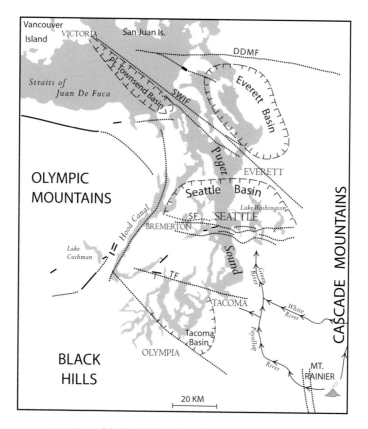

FIGURE 8.6 Crustal faults of Puget Sound region, showing Seattle fault (SF) between Bremerton and Seattle. Well-located active faults in solid heavy lines. Other faults (dotted lines) include Tacoma fault (TF) and the Southern Whidbey Island fault (SWIF), which extends from offshore Victoria to the mainland south of Everett and is the site of a regional wastewater disposal plant. Short dashes are in the direction of fault dip. Green River, White River, and Puyallup River (arrows) are potential courses of volcanic mudflow deposits from Mt. Rainier. Source: based on work led by Brian Sherrod of USGS and Tim Walsh of Washington Dept. of Geology and Earth Resources. Modified from Yeats (2004, Fig. 6.9).

of at least 7. This earthquake is also mentioned in Native American oral histories. The Portland Hills fault is found on both sides of the Portland Hills in downtown Portland (Figure 8.9), although the evidence for Holocene rupture is not as clear in Portland as it is in Seattle. The affected area of the next earthquake in the Puget Sound

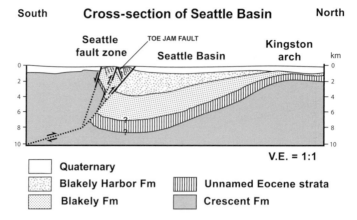

Cross-section of Seattle Basin

South ... North

Seattle fault zone   TOE JAM FAULT   Seattle Basin   Kingston arch   km

V.E. = 1:1

Quaternary
Blakely Harbor Fm
Blakely Fm
Unnamed Eocene strata
Crescent Fm

FIGURE 8.7 North–south cross-section of the Seattle area, locating the Seattle crustal fault, which is mapped (Figure 8.6) from Bainbridge Island across Puget Sound and downtown Seattle. Source: cross-section based on seismic profiling led by Tom Brocher of USGS.

FIGURE 8.8 LiDAR (LIght Detection And Ranging) image of Toe Jam Hill fault on Bainbridge Island (straight trace, with south-facing fault scarp, in shadow). The image illuminates the ground with a laser and analyzes the reflected light to produce a detailed bare-ground topographic image. In this image, light source is from northwest, elevated 40° above the horizon. The fault is at right angles to glacial grooves. Waters of Puget Sound unshaded; note marine terrace just above the shoreline. Source: image created by Brian Sherrod, USGS.

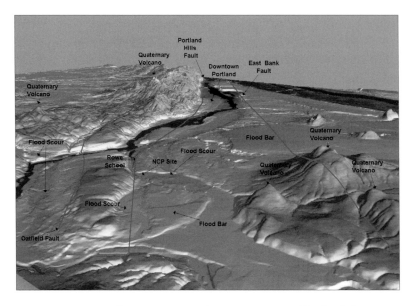

FIGURE 8.9 Computer-generated perspective map of Portland, Oregon, showing zone of faulting on both sides of Portland Hills and volcanoes, which are not known to be active. Vertical scale is exaggerated. Topography also reflects scouring by deposits of the Missoula Floods in latest Pleistocene time. Faults are shown in solid lines. The Portland Hills fault has been located at Rowe School, although it is not known to cut Holocene deposits there. Source: image from Ian Madin, Oregon Department of Geology and Mineral Industries.

or Portland would be smaller than on the subduction zone, but because the faults are closer to the surface the damage in local areas could be intense.

OUR CHOICE

The lesson of the resilience studies of Oregon and Washington is that there is a monstrous price tag in doing nothing or in taking only token steps. Failure to strengthen lifelines, including bridges and utility networks, city halls, police stations, school buildings, and hospitals, has a huge cost: a decades-long decline of the Northwest economy. If industry is forced to leave the Northwest, then one must factor in sharply lower tax revenues and sharply higher unemployment. This will require outside assistance to the Northwest and British Columbia from the national governments of the United States and Canada. Both countries would be required to play a major financial role in recovery and, for the US government, would defer the reduction of the national debt, currently one of its highest political priorities.

The federal government has taken a large role in recovery from Katrina and is expected to take a major role in the recovery from Sandy, at least in the most devastated coastal regions of the states of New York and New Jersey. A bill in Congress to provide more than $60 billion for recovery in New York and New Jersey failed to pass due to Congressional dithering about the fiscal cliff. The bill was only passed months later due to the public outrage against Congress for failing to take action. It was not an encouraging sign about how the federal government would respond to the Cascadia earthquake.

Port-au-Prince, largely destroyed by the 2010 earthquake, still has not recovered despite international assistance. Up to 400,000 people are still living in refugee camps. The presidential palace and many commercial buildings have not been rebuilt. A lot of bricks are being made, but the buildings made from them will probably be unreinforced masonry, a recipe for collapse in a future earthquake.

A more realistic comparison of the destruction from the next Cascadia earthquake is the firebombing of Japanese cities in the late stages of World War II or the devastation of the economy of the southern United States caused by the Civil War. These were disasters affecting the economy on a national scale. In each case, recovery took decades, at least a generation.

Unfortunately, science cannot tell whether the Cascadia earthquake will strike tomorrow or a century from now, but science can say, without a doubt, that *there will be an earthquake as large as magnitude 9 in the near future.* If the paradigm shift about the next Cascadia earthquake is to be realized among the general population of the Northwest, action must begin *now.* This means that the forthcoming earthquake needs to be debated and dealt with by the state and provincial legislatures, by civic leaders, by the media, and by the public at large, because it will require taxpayer dollars, including assistance from the federal government. It will require a financial commitment by society, both at the state and provincial level and at the national level, to prepare for the next earthquake *now* so that when it strikes, it will be a manageable crisis. A commission authorized by the Oregon legislature (Governor's Task Force on Implementation of the Resilience Plan, authorized by SB 33 and chaired by Dean Scott Ashford of the Oregon State University College of Engineering) recommended an expenditure of $100 million per year to prepare for the next subduction zone earthquake, in hopes that the earthquake does not strike before many of the recommended actions have been taken. At the present time, the Oregon resilience survey has not led to any action by the Oregon legislature, although ex-Governor Kitzhaber of Oregon included funds in the 2015 budget he presented to the State legislature. But, as stated above, the legislature took no action to repair Oregon's obsolete bridges.

The question is: who pays?

The Washington State resilience survey contains recommendations on responding to the earthquake threat in the next 50 years. The actions to be taken include response by the office of the governor and the state legislature, which will in turn require a sacrifice by the residents of the state. This involvement assumes that the governor and legislature will have the political will to act on the resilience survey, including providing the necessary resources. It also assumes that the citizens will recognize what needs to be done *in advance* of the earthquake and will demand action by their elected leaders. Will they act? The same question may be asked in Oregon, northern California, and British Columbia.

As a resident of the Northwest, I find the failure to take meaningful action against this threat unacceptable and irresponsible. Will we and our elected officials rise to the challenge? Can my colleagues and I make a more convincing case leading to action?

My colleague at the University of Tokyo, Professor Yasutaka Ikeda, warned his countrymen in advance that the Tohoku region of northeast Japan should be preparing for a subduction zone earthquake of magnitude 9, not 8–8.4 as had been estimated previously. Failure to prepare for a magnitude 9 earthquake meant that the tsunami-protective seawalls were not high enough and, as a result, nearly 16,000 people lost their lives. After the earthquake, I asked Ikeda-san if he was telling his countrymen "I told you so." He sadly said no, that he blamed himself for not doing a better job of convincing the Japanese authorities. To him, the inability to make the case to his countrymen for a magnitude 9 in advance was the greatest failure of his long and distinguished career.

I have similar concerns.

The Oregon and Washington resilience surveys laid out a plan to strengthen the states against the inevitable Cascadia earthquake over the next 50 years. If we start now, there is a probabilistic 90% chance we can strengthen ourselves against the next earthquake before it happens, if our probability estimates are correct. But at the present time, this does not appear to be on the legislature's to-do list in either state. No bills have been introduced, and no attempt has been made by state politicians to raise the money, although, as noted above, there is a Governor's Task Force chaired by Dean Scott Ashford to consider Oregon's response to the problem, authorized by Senate Bill 33. But the lack of outrage by the general public that state officials have not yet addressed the threat of the next Cascadia earthquake is evidence that the paradigm shift has not yet occurred among the general public.

However, Steve Novick, a Portland city commissioner, has taken notice, and on October 16, 2013, he proposed in the Portland *Mercury* a policy of mandatory seismic retrofits for URM buildings in the city of Portland, a policy favored by Portland emergency managers. This would reverse a weakening of the seismic code for Portland that had been agreed to unanimously by city commissioners in 2004 in response to complaints by developers and apartment owners.

There are about 1300 URM buildings in Portland, and more than 5000 across Oregon, mostly constructed before about 1930. Some of the building owners do not live in Portland and thus have no investment in the city other than their own URM building.

What was the response of URM owners and landlords in Portland to Novick's proposal? They formed the Heritage Bricker Housing and Jobs Coalition, abbreviated Brickers, and registered as Masonry

Building Owners of Oregon, a lobbying group that has contributed large checks to current commissioners and to Mayor Charlie Hales. About 25 Portland landlords own 125 unsound brick apartment buildings, and this organization claims that they cannot afford to retrofit those buildings on their own. Their solution: Pay us, and we'll retrofit our own URM buildings. There was no mention in the *Mercury* article about concern among the URM building owners that a subduction zone earthquake could kill hundreds of people in those buildings, or what the source of the retrofit funding would be. So the debate over retrofitting dangerous buildings in Portland was foundering over the question of who pays.

Enter the Historic Preservation League of Oregon. This organization, based on their roundtable in 2012, has determined that some of the most important historic buildings in Oregon are at greatest risk of earthquake collapse. One of their findings is that the upper floors of some of the URM buildings are already uninhabitable, meaning that such buildings are already reduced in the rental income that they can generate. Government, from federal agencies to local jurisdictions, has a responsibility to upgrade those buildings for which they are responsible. Two earthquakes in Oregon in 1993 produced major damage to government buildings, including $4.5 million in damage to the State Capitol in Salem and more than $3 million in damage to the Klamath County Courthouse.

One way to reward seismic upgrades is to modernize rehabilitation tax credits, which would create jobs. From 1978 to 2011, rehabilitation tax credit projects totaled $116.5 billion. These projects generated 2.2 million jobs and rehabilitated 42,000 structures, large and small, around the state. Tax credits of $19.2 million generated $24.4 billion in new federal tax receipts and $9.1 billion in increased state and local revenues for smaller projects. Congressman Earl Blumenauer is on record as favoring an update of the federal tax credit program.

Buildings now have a rating system called LEED (Leadership in Energy and Environmental Design). A similar certification system should be developed for seismic upgrades. This is already being done through the Structural Engineers Association of northern California.

Urban renewal projects could borrow against the increased tax revenue for the lifetime of the project, commonly 15 years. The Northwest also needs an earthquake insurance pool for historic properties. California has the California Earthquake Authority, which serves this purpose in the California part of Cascadia, as well as the rest of the state.

In conclusion, the goal of making buildings safe against the next inevitable Cascadia earthquake starts with a large dollar figure, but by government taking the lead and undertaking rehabilitation in segments, as well as providing incentives to building owners through taxation modification and publicizing buildings that have been modified and are now safer to use, the Pacific Northwest can get started on eliminating its URM inventory, hoping that the earthquake doesn't strike first.

I appreciate the insights provided by an investigative journalist, Nathan Gilles, who wrote one of the key articles in the Portland *Mercury*, and by Scott Ashford, Dean of the OSU College of Engineering.

STEPS TAKEN IN THE CALIFORNIA PART OF CASCADIA AND IN BRITISH COLUMBIA

The southern part of the subduction zone is in California, which experienced a small subduction zone earthquake in 1991. Since 1986, state legislation has required local jurisdictions in harm's way from earthquakes to conduct inventories of hazardous URM buildings and to remove these buildings from use. The City of Eureka has had a program since 1989, and most of the buildings identified in Eureka's inventory have now been retrofitted. No comparable program exists in Oregon or Washington, and in Portland, where the debate between Novick and Brickers is taking place, it is unclear how bad the URM problem is, in contrast to the California part of Cascadia, where an inventory has already taken place. In addition, in 2003, the City of Vancouver, BC, installed a dedicated fire-protection system that will allow firefighters to pump water from two nearby creeks, important because of the danger of fire after a major urban earthquake.

Small but important steps.

9　Japan: Tokyo and the Kansai

Japan is the best-prepared nation on Earth against earthquakes. Earthquakes are part of Japan's cultural heritage and have even been the subjects of classical Japanese art. The first catalog of Japanese earthquakes was created in AD 900. Two subduction zone earthquakes in 1854 did great damage in the western part of the country (Figure 9.1), especially in the region between the two great population centers of Tokyo and Kyoto–Osaka–Kobe, the latter region called the Kansai.

However, by the time of the Meiji Restoration in 1868, the Japanese had decided to not just endure earthquakes and tsunamis but to do something about them. They hired foreign experts to study earthquakes, with their research centered at the Imperial University of Tokyo. The world's first professional society dedicated to the study of earthquakes, the Seismological Society of Japan, was founded in 1880; more than half of its members were foreigners. Public interest increased dramatically in 1891 after an earthquake of magnitude 8 struck western Japan north of Nagoya, with the loss of 4000 lives (Figure 9.1). Investigation of this earthquake was led by Professor Bunjiro Koto, a geologist at the University of Tokyo. Another expert was Professor Fusakichi Omori, a seismologist who, at the time of the 1906 San Francisco earthquake, was the world's leading earthquake seismologist. He participated in the Carnegie Institution of Washington-funded investigation of that earthquake.

Japan has two strikes against it. First, it is bounded by not one but two subduction zones, one extending north of Tokyo up the east coast of Honshu and Hokkaido, and the Nankai subduction zone located in southwest Japan south of the Kansai region (inset, Figure 9.1). Second, Japan is a small, crowded country, and less heavily populated regions are limited to the northern Home Island of Hokkaido.

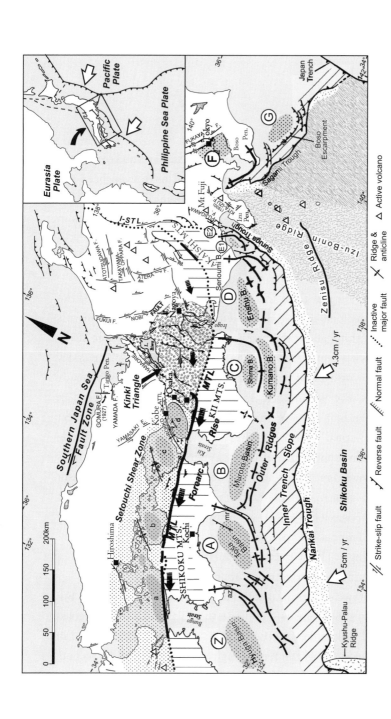

FIGURE 9.1 Active tectonic map of southwest Japan. Two earthquake time bombs, the Tokyo–Yokohama area (f) and the Kansai, including Osaka, Kyoto, Nara, and Kobe, have generated damaging earthquakes. A fault west of Osaka (between (c) and (d)) was reactivated in 1995 in the Kobe earthquake. Mt. Fuji (open triangle north of Suruga trough), an active volcano, is located northeast of the Suruga Trough, site of an earthquake of magnitude 6.2 in 2011. The great Kanto earthquake of 1923 was on the subduction zone marked by Sagami trough. An additional crustal hazard to Tokyo is formed by several faults cutting through the city (area F). MTL = Median Tectonic Line, an active strike-slip fault. Source: after Sugiyama (1994), Yeats (2012, Fig. 9.16).

By the early 1920s, the Japanese scientific community was taking initial steps to understand and deal with the earthquake problem, although the Japanese government and the general public were not fully involved. This situation changed suddenly on September 1, 1923, when the Tokyo–Yokohama metropolitan area (Figure 9.1) was struck by an earthquake of magnitude 7.9, resulting in 142,800 deaths and an additional 40,000 missing. The earthquake struck just before noon, when families were preparing their midday meal on charcoal stoves; the strong shaking caused fires all over the city.

World War II delayed the scientific response to the Tokyo–Yokohama earthquake, but in 1944, while the war still raged, the Nankai subduction zone off southwest Japan was struck by a subduction zone earthquake of magnitude greater than 8, in which 1251 people were killed (Figure 9.1, C and D). A tsunami with wave heights as great as 30 feet swept over the coast.

After the 1944 earthquake, Akitune Imamura, one of the pioneers in earthquake studies, warned that the western half of the Nankai subduction zone (Figure 9.1, A and B) had not yet ruptured, although it had done so in previous earthquakes documented over the preceding 13 centuries. Imamura wrote a letter to the Academy of Japan in October 1946, warning of the likelihood of another earthquake on the western part of the subduction zone in the very near future. Nothing was done, and the 1946 Nankaido earthquake, also with magnitude greater than 8, struck the western part of the Nankai subduction zone. Imamura's warning had not been heeded, and he died in obscurity in January 1948.

Chastened by failing to forecast the Nankaido earthquake, the Japanese establishment focused on their apparent understanding of the different distribution of the 1944 earthquake as compared with the previous earthquake in that area in 1854. The 1854 earthquake apparently had been documented in surface ruptures in the floodplain of the Fuji River north of Suruga Bay (Figure 9.1, E2), but the 1944 earthquake had not extended that far east. Accordingly, the Japanese focused on the different extent of the two earthquakes to define a feature they called the Tokai Seismic Gap, in which the easternmost part of the 1854 rupture had not ruptured in 1944 and was thus a highly likely site for another earthquake, posing a hazard to the cities in the "gap." The Japanese government supported detailed instrumentation of the Tokai Gap with the objective of "capturing" the expected earthquake.

Based on a project my consulting firm did in the region after a damaging earthquake of magnitude 6.2 in 2011 in Suruga Bay, south of the surface ruptures in the Fuji River, we concluded that the 1854 surface ruptures in the Fuji River could have accompanied a local earthquake, not necessarily on the Nankai subduction zone. If that were the case, then there would be no subduction zone gap and no reason to focus earthquake resources on the Tokai region.

Indeed, the next earthquake to strike southwest Japan occurred on January 17, 1995 in the city of Kobe, farther west (fault between C and D, Figure 9.1). The port city of Kobe at the time was not considered to be an urban region requiring a high priority for retrofitting. Kobe was severely damaged by an earthquake of magnitude 6.9 in 1995, incurring losses of $150 billion, the failure of 80% of the city's small businesses, the shutdown of the chemical and steel industries for several months, the loss of gas service to 845,000 households for two months, and the loss of water and sewage treatment for four months. The shutdown of Kobe port, then the sixth largest port in the world, diverted port business to Yokohama, Osaka, and South Korea. After three years, 10–15% of commercial business at Kobe port had not returned, and the world ranking of Kobe port had fallen from 6th to 39th place. The recovery of Kobe depended largely on major assistance from the Japanese government, which also responded to the Tohoku tsunami of 2011.

In hindsight, Kobe was an earthquake time bomb. This was not generally recognized at the time of an international meeting of earthquake geologists in Kobe that I attended in 1984. That meeting was followed by a field trip to the Rokko Mountains west of Kobe, led by Kazuo Hujita, a professor at Osaka City University. Professor Hujita pointed out evidence of active faulting in the Rokko Mountains that presented a major earthquake threat to the city of Kobe. Despite his warning, the investigations on earthquake hazards continued to focus elsewhere in Japan. The 1995 earthquake, like the 2010 Port-au-Prince earthquake, caused great damage and loss of life. Among the casualties was Hujita's classical Japanese home in Kobe.

In recognition of his foresight, this section on Japan is dedicated to Professor Hujita's memory.

The Tohoku area had been recognized as subject to earthquake hazard, critical because of the presence of nuclear power plants along the coast, operated by the Tokyo Electric Power Co. (TEPCO). The power company had considered earthquake hazards based on the seismic history of the subduction zone over the past century or two. Based on this history, TEPCO estimated the size of the maximum

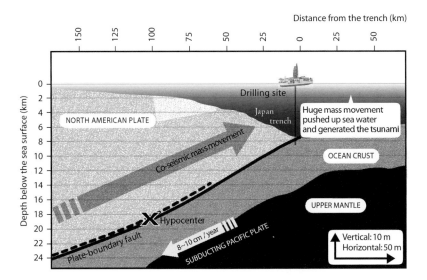

FIGURE 9.2 Cross-section of subduction zone fault generating the 2011 Tohoku-oki earthquake of magnitude 9. Expected rupture zone of a subduction zone earthquake prior to 2011 is shown in heavy dashed line. The actual rupture zone, in heavy solid line, was close to the Japan trench, contributing to the size of the tsunami that killed nearly 16,000 people. The ship is the *Chikyu*, which obtained cores from the fault, enabling the measurement of frictional heat released during the earthquake by scientists at University of California at Santa Cruz. Source: modified from Lin *et al.* (2013).

earthquake to strike Tohoku as magnitude 8.4. Instead, on March 11, 2011, the subduction zone was struck by an earthquake of magnitude 9, the largest to strike Japan in its long recorded history.

Not everyone in Japan was caught by surprise. Professor Yasutaka Ikeda of the University of Tokyo, using GPS data as well as evidence for the shortening rate across northeastern Japan, estimated in 2010 that the Tohoku region could be struck by an earthquake of magnitude 9. Koji Minoura, a geologist at Tohoko University in Sendai, determined in 2001 that the Sendai region had been struck by a major tsunami in AD 869, based on tsunami deposits found by his team as far inland as the deposits from the 2011 tsunami. The Japanese scientific establishment and TEPCO failed to consider these opposing views, and a catastrophe was born, not only the drowning of nearly 16,000 people in the tsunami (Figure 9.2), but also the radioactive releases at the Fukushima Dai-ichi nuclear power plant.

Recent work, including drilling by the Japanese research vessel *Chikyu*, gives two reasons for the unexpected large size of the tsunami: an earthquake magnitude larger than expected, and a location of the surface rupture very close to the Japan trench, similar to that of the 1896 Sanriku earthquake and tsunami, in which the tsunami also did an unexpected amount of damage.

Japan is, indeed, the best-prepared nation on Earth against earthquakes. But in not considering alternative ideas being advanced by some of their own scientists, they prepared for the wrong earthquake.

10 Wellington, New Zealand

In 1983, I received a grant from the National Science Foundation to spend a year in New Zealand working with the New Zealand Geological Survey to better understand reverse faults affecting new hydroelectric power projects in the South Island. New Zealand geologists understood very well the hazard from strike-slip faults, and I had experience with reverse faults in southern California. My New Zealand counterpart was Kelvin Berryman, who put me to work on a fault impacting a dam proposed by the Ministry of Works and Development.

I quickly found that the Kiwis are quick learners, and I found myself learning as much from them as they learned from me. I also found that New Zealanders are as aware of earthquake hazards as my California colleagues, and their awareness includes strong building codes and a strong background in earthquake studies, not only at the Geological Survey offices north of Wellington but at the universities, including Victoria University of Wellington.

The reason for their awareness is found in their earthquake history. New Zealand was struck by major earthquakes in 1848, 1855, 1888, 1929, and 1931. The 1855 West Wairarapa earthquake on a strike-slip fault east of Wellington had a magnitude of 8.2, larger than any earthquakes in the past 150 years. Sir Charles Lyell, one of the founders of active fault studies, wrote about this earthquake in the twelfth edition of his textbook, *Principles of Geology*. The surface fault rupture associated with this earthquake is very well exposed in the Wairarapa Valley, east of the capital city of Wellington, and over the years many geologists have taken field trips to view this fault.

The country was thinly populated during most of the period prior to World War II, and so losses from these earthquakes, although locally severe, did not affect the economy of the nation. But during the summer of 1942, the Wairarapa Valley was struck by two earthquakes, the largest of magnitude 7.2, and hundreds of homes were severely damaged. The earthquakes took place during the darkest days of World War II, which was being fought in islands to the north, and there was little money available for reconstruction. In 1944, while the war was still going on, Parliament passed the Earthquake and War Damage Act, and in January 1945 the government began collecting a surcharge from all holders of fire insurance. A government commission was established to collect premiums and pay out claims for damages from war or earthquakes. Later, coverage against volcanic eruptions, tsunamis, and landslides was added.

In 1988, Parliament changed the governing body from a government organization with an insurance commissioner to a corporation responsible for its own funding, paying a fee to the government as a guarantee in case a natural disaster was so large that the insurance reserve fund was exhausted. In 1993, the governing body was renamed the Earthquake Commission (EQC). The EQC automatically covers against earthquakes all residential properties insured against fire, up to NZ$100,000. The arrangement has worked well since 1944, in large part because New Zealand has not experienced an earthquake as large as the 1855 earthquake since the fund was established. The cost of damage from an earthquake of magnitude 7.8 in 1931 at Hawke's Bay on the North Island was NZ$850 million in 2010 dollars. Costs of recent earthquakes have increased from NZ$2.4 million from the 1968 Inangahua earthquake in the South Island to NZ$136 million from the smaller 1987 Edgecumbe earthquake in the North Island. The earthquake fund accumulated reserves to pay out in future earthquakes, and part of the fund was used to buy reinsurance on the global market.

This system was put to the test when an earthquake struck in an unexpected place in Canterbury in the South Island: Christchurch, New Zealand's second largest city, which had been considered in a moderate- to low-hazard zone. An earthquake of magnitude 7.1 struck a previously unknown fault 25 miles (40 km) west of Christchurch on September 4, 2010, resulting in damages of NZ$2.75 to NZ$3.5 billion, but no deaths. This was followed in February 2011 by a second earthquake of magnitude 6.3 within the city, taking 185 lives and causing massive destruction (Figure 10.1). Further damaging aftershocks continued through the rest of 2011. The second earthquake was smaller, but because it struck within the city of Christchurch itself, the losses were much higher. The initial estimate of cost was NZ$15 billion, but this estimate was raised to NZ$40 billion by April 2013. Payout for losses came from several sources: (1) the international reinsurance sector, which covered the bulk of the losses; (2) the earthquake fund that had been accumulating through the EQC for decades; (3) separate private insurance; and (4) direct government assistance. As a result, the cost of the Christchurch earthquakes, the most costly natural disaster in New Zealand's history, was

FIGURE 10.1 An earthquake in February 2011 in New Zealand's second largest city, Christchurch, produced major damage and loss of life. Prior to this earthquake and an earlier one in 2010, Christchurch had not been considered at major risk from earthquakes. Source: Sharon Davis through Flickr.

manageable, although some of the 500,000 claims for residential damage alone are still being processed.

It has helped that the people of New Zealand, like California, Japan, and Chile, are aware of their earthquake hazard and have prepared themselves accordingly. Because of reinsurance, money actually flowed *into* the country after the earthquake to pay claims.

However, there are concerns about three different earthquake sources. The 1931 Hawke's Bay earthquake of magnitude 7.8 occurred near or on the Hikurangi subduction zone between the Pacific and Australian plates. The greater-than-expected magnitude of the Tohoku earthquake in Japan raises the question of how reliable the estimates are of the maximum earthquake size on the Hikurangi subduction zone. Could the subduction zone generate a much larger earthquake than it did in 1931? This region is a major focus of research; the most recent hazard model has a magnitude 9 earthquake as a possible maximum likelihood, with accompanying major displacement and large tsunami.

A second concern is the Wairarapa fault, which generated an earthquake of magnitude 8.2 in 1855. The strike-slip rate on the Wairarapa fault is 7–10 mm per yr (¼–⅓ inches per year). The earthquakes in 1942 on a different fault in the Wairarapa Valley indicate that a repeat of these earthquakes or still-larger earthquakes could produce major damage in the Wellington region, with a current population of 400,000.

Finally, Wellington has its own strike-slip fault that extends through the downtown area, close to Parliament buildings and the major commercial districts. Most of the population lives within a few miles of the Wellington fault (Figure 10.2). Paleoseismological studies show that the slip rate on the Wellington fault is 5–7 mm per yr (about ¼ inch per year). The last earthquake occurred within the past 300 years, and the preceding earthquake struck 800–900 years ago. Kelvin Berryman has estimated that the past major earthquakes on the Wellington fault have been in the range of magnitude 7.3–7.9.

Berryman now manages natural hazards research for the Government of New Zealand through GNS Science, the successor to the

FIGURE 10.2 Aerial view of Wellington fault, view toward south. Its straight trace (arrows) is due to its main displacement being strike-slip. At far left, fault extends through the capital city of Wellington (arrow on far left) close to Lambton Harbour, extending north as a scarp along the west side of the Hutt Valley. Across Cook Strait in the far distance are snow-capped mountains of the South Island. Source: photo courtesy of New Zealand Institute of Geological and Nuclear Sciences.

New Zealand Geological Survey. He and his team responded to the Christchurch earthquakes of 2010–2011, the most devastating earthquakes in the 200-year recorded history of that country and the largest socio-economic crisis to affect New Zealand since World War II. The damage from the Christchurch earthquakes was more than NZ$40 billion, equivalent to 15% of New Zealand's gross national product.

Yet the effect on the national economy was minimized because more than 80% of properties are covered by insurance, managed through New Zealand's Earthquake Commission for residential properties and through the open market for commercial properties. Local insurance is underwritten by policies through overseas reinsurance companies. Nevertheless, the New Zealand government will pay about $15 billion of the bill for recovery, and the rest will be paid by local insurance and reinsurance providers. The EQC's earthquake fund was depleted by the Christchurch earthquakes, but the EQC is now beginning to accumulate reserve funds. For all New Zealanders, the hope is that the next earthquake does not strike until that fund is replenished. In the meantime, one effect of the Christchurch earthquakes has been a substantial increase in earthquake premiums. A review of the EQC Act is in progress.

11 Santiago, Chile

On February 27, 2010, the subduction zone off southern Chile was the source of the Maule earthquake of magnitude 8.8 in which 525 people were killed, including 23 in metropolitan Santiago (Metropolitana), with an additional 25 missing. Damage estimated by the reinsurance industry was $4–7 billion, and the cost to the national economy was $15–20 billion. These are large numbers until one considers that this was the sixth largest earthquake ever recorded by a seismograph. The losses were large, but considering that Maule was a superquake, the outcome could have been much worse.

Chile's response to the Maule earthquake was influenced by the fact that Chile has been struck by earthquakes on the Peru–Chile subduction zone many times during its recorded history of nearly 500 years. An earthquake of magnitude 9.5 on May 22, 1960 in southern Chile was the largest earthquake ever recorded on seismographs. At the time of the Maule earthquake, the government and the general population were aware of earthquakes, and, in the wake of the great 1960 earthquake, they had upgraded their building codes and zoning regulations. The losses of life accounted for only 0.4% of the population in the area affected by strong ground motion, a disaster but not a catastrophe.

The earthquake struck at the end of the summer holidays, and many of the casualties were tourists from other countries, who were completely unaware that Chile has an earthquake problem. Because Chile is a developed country, the Pacific Northwest can learn important lessons from the Maule earthquake, as pointed out in the resilience surveys summarized by CREW (2013).

Santiago, the capital of Chile, has a population of nearly six million, nearly one-third of the population of the entire country.

The epicenter of the 1960 earthquake was 350 miles (570 km) south of Santiago, and the epicenter of the 2010 Maule earthquake was closer to Santiago: 210 miles (335 km) south of the capital city (Figure 11.1). There were damage and losses in the Santiago Metropolitana region from the Maule earthquake, but these were manageable. Santiago's test will come with a subduction zone megaquake directly opposite Santiago itself.

Like the cities of the Pacific Northwest, Santiago faces earthquakes from the South American crust in addition to earthquakes on the subduction zone. The two largest faults, both strike-slip, are the Atacama fault of northern Chile and the Liquiñe-Ofqui fault of southern Chile. These faults are too far away to be a hazard to Santiago. But Rolando Armijo, a Chilean expat now at the Institut de Physique du Globe de Paris, has returned to his native land to study crustal faults in the Santiago metro region. Armijo has shown that the boundary between the Central Depression, where Santiago is located, and the West Andean mountains to the east is a major reverse fault called the West Andean thrust (Figure 11.2), which includes the San Ramón thrust. A fault scarp formed by the San Ramón thrust is 11–13 feet (3–3.7 meters) high; this could have been formed by a prehistoric earthquake of magnitude 6.6–7.

A swarm of earthquakes with magnitudes as large as 6.3 struck over a period of six minutes in 1958. Most of Santiago was destroyed by an earthquake in 1647 that might have struck a crustal fault rather than the subduction zone. Armijo and his colleagues, including Chilean geologists at the University of Santiago, estimate that the slip rate across the Andes at the latitude of Santiago is low enough that the recurrence interval on Andean crustal earthquakes could be as long as 2500–10,000 years.

Chile's success in responding to recent earthquakes appears to be related to its experience of past earthquakes, including the 1960 superquake. Armijo and his Chilean colleagues have now embarked on a paleoseismic program to determine the earthquake history of the San Ramón thrust and other structures in the West Andean fault zone.

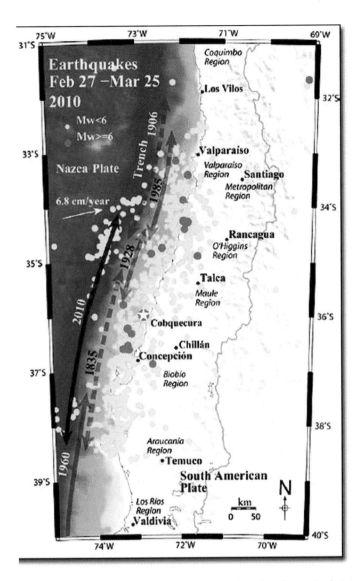

FIGURE 11.1 2010 Maule subduction zone earthquake of magnitude 8.8 compared with location of Santiago metropolitan region. Solid heavy line with arrows shows the rupture zone of Maule earthquake; to the south is the north end of the rupture zone of 1960, which overlaps the 2010 rupture zone by about 100 km. The dashed line marks rupture zone of 1835 earthquake experienced by Charles Darwin. Rupture zones of smaller subduction zone earthquakes opposite Santiago and Valparaiso also shown. Dark and light circles identify earthquakes of the 2010 Maule sequence. Permission to use image from sms-tsunami-warning.com.

FIGURE 11.2 View from Santiago, east to snow-capped Andes. Boundary between city and mountains is marked by the West Andean thrust, which may be a crustal threat to the city. Source: photos © Pablo Viojo/Flickr, made available under an Attribution 2.0 Generic (cc by 2.0) license at www.flickr.com/photos/pviojoenchile/2733293472.

These are still predominantly projects funded by research contracts, although, as I have stated for other earthquake time bombs, the large population of Santiago at risk from earthquakes is a strong argument for direct government funding.

12 Prologue in central China

The explosive migration of humanity to major cities in the developing world is by and large a modern phenomenon, a byproduct of the growth of population worldwide in the past century. Most of the time bomb cities featured in this book either did not exist at all hundreds of years ago, like Port-au-Prince or Nairobi, or were relatively small cities at that time, like Kabul. There is one major exception, however; the cities of central China.

As discussed below, China has the largest government organization on Earth for the study of earthquakes, and so its cities are as well prepared as those discussed above. The difference in central China is that a population explosion took place hundreds of years before others discussed below, and so losses were high, with one earthquake causing more deaths than any others in Earth's history. For this reason, this chapter is described as a "prologue."

China's civilization is thousands of years old, and during most of that time its capital was in the valley of the Wei River (*Weihe*), itself a tributary of the Yellow River (*Huanghe*) in modern Shaanxi Province (Figure 12.1). The famous terra cotta warriors of modern Xi'an were an inspiration of Qin Shi Huangdi, the first emperor of China, in the third century BC. During the Han Dynasty, 2000 years ago, the city of Chang'an in the Wei River valley was the Chinese terminus of the Silk Road between Europe and the Far East. For many centuries, Chang'an was a center of Buddhist and Taoist culture, with great temples and pagodas. Even Christians thrived there. Chang'an, renamed Xi'an in the fourteenth century AD, had streets that were laid out on a grid pattern oriented north–south and east–west, with broad avenues, some with median strips. The street plan became a model for other Asian cities, including Nara and Kyoto in Japan, and is followed by many American cities today, such as Salt Lake City, Utah.

In medieval times, the city of Chang'an and the valley of the Wei River attracted hundreds of thousands of migrants. They moved there to be part of one of the most vibrant cities on Earth. The city had been founded in 195 BC, and it soon had a population of nearly 150,000. By AD 750, Chang'an had grown to a population of between

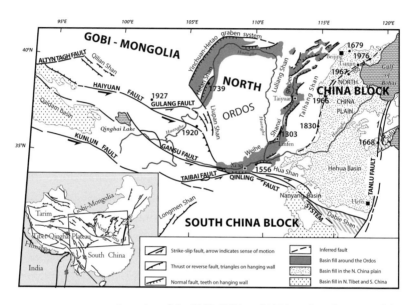

FIGURE 12.1 Location of the 1303, 1556, and 1739 earthquakes around the Ordos Plateau in central China, and the 1668 Tanlu, 1679 Sanhe, and 1976 Tangshan earthquakes east of Beijing. All are located on or close to active faults, and all are estimated to be as large as magnitude 7.5–8. Other earthquakes: 1920 Haiyuan fault earthquake of magnitude 7.8 on the Haiyuan fault, with loss of life of more than 200,000 according to the USGS, and 2008 Wenchuan earthquake of magnitude 7.9 on faults in the Longmen Shan, in which more than 87,000 people lost their lives (epicenter not shown). None of the earthquakes struck close to a plate boundary. Dashed line: Tangshan–Hejian–Cixian fault named from earthquakes in 1976, 1967, and 1830; 1966 Xingtai earthquake also on this fault. TSG, Tianjin seismic gap between 1967 and 1976 earthquakes identifies Tianjin as an earthquake time bomb. Source: modified from Yeats (2012, Fig. 8.16) and Zhang *et al.* (1998), with permission granted from Elsevier.

800,000 and one million, and in the next 200 years it was probably the largest city on Earth, with a population of nearly two million in the metropolitan area. The poet Ban Gu wrote about the crowds in the Chang'an bazaars:

> There was no room for people to turn their heads,
> Or for chariots to wheel about.
> People crammed into the city, spilled into the suburbs,
> Everywhere streaming into the hundreds of shops.

The city was the capital of China during more than ten dynasties, and, even after the capital moved out of the region in the fourteenth century, the city continued to thrive. Just as the city had grown, surrounding regions also attracted many migrants, moving there for reasons similar to migrations to major cities today. Housing for new arrivals took advantage of a peculiar feature of the landscape of central China, the presence of great thicknesses of wind-blown silt, called *loess*. The region north of the Wei River valley, the Ordos Plateau, was dominated by hills underlain by loess, and the loess hills extended south to Chang'an (Figure 12.2). (Centuries later, the city of Yan'an in the Ordos Plateau would be the destination of the Chinese Communist Long March in the 1930s.) People found that they could excavate dwellings in the soft loess silt, and so hundreds of thousands of them lived like swallows in man-made caves called *yaodongs* carved into the loess hills.

Catastrophe lurked in earthquake faults that cut through the loess hills, faults that were completely unknown to the emperors who ruled the region in medieval times. Unlike most of the faults posing hazards to earthquake time bombs in this book, these faults are not on plate boundaries, and so even modern geologists find it hard to determine their significance in earthquake hazards. The first fault in many centuries to rupture and reveal itself to an unsuspecting population more than 700 years ago was not in Chang'an, but hundreds of miles away to the northeast, near the city of Linfen in the modern province of Shanxi, the province name meaning "west of the [Taihang]

FIGURE 12.2 Loess hills of Qishan County, Baoji City, Shaanxi Province, where thousands of people have excavated their homes within the loess (*yaodongs*). Similar yaodongs collapsed during the earthquake of 1556, entombing the inhabitants and resulting in a death toll greater than 800,000, the largest death toll from an earthquake in recorded history. Source: photo taken by Li Gaoyang of Earthquake Administration of Shaanxi Province, provided courtesy of Ren Junjie, China Earthquake Administration, Beijing.

Mountains." Like Chang'an, Linfen was an ancient city. Legends held that it was founded more than 11,000 years ago, and for a time it was known as China's first capital. It now has a population greater than 3.5 million, but was already a large city at the time of a huge earthquake of magnitude 8 on September 17, 1303, during the Yuan dynasty (Figure 12.1). The Hongtong earthquake took the lives of more than 200,000 people, many killed by landslides shaken loose by the earthquake.

But the worst was yet to come.

On January 23, 1556, during the reign of Emperor Jiajing of the Ming dynasty, a massive earthquake, also of magnitude 8, struck close to the densely populated city of Chang'an/Xi'an, killing 830,000 people, the greatest loss of life from an earthquake in human history. Buildings in an area 520 miles wide were destroyed, and 60% of the people in the area of worst damage were killed. Great fires raged for several days. Some of the greatest losses were to the millions of people living in yaodongs excavated in the loess, which became death traps that collapsed and entombed them, in part triggered by earthquake-activated landslides that buried the entrances to the caves (Figure 12.2). A scholar named Qin Keda provided one of the first descriptions of an earthquake disaster, including the destruction of the yaodongs and heavy damage to cultural features in the city, including stelas of stone containing inscriptions (Forest of Stone). Many were destroyed. Qin Keda even provided advice on surviving an earthquake: Stay where you are, and don't run outside. This is still good advice even today.

The earthquakes, including another one that offset the Great Wall of China in 1739, and another one near Beijing in 1679, was an example of *earthquake clustering* (Figure 12.1). They struck different faults, but for a particular region their return times were much shorter than would have been expected considering individual earthquakes and faults separately.

For centuries, the Chinese simply endured the frequent earthquakes that struck their cities, but finally, during the modern era, they had had enough. The philosophical origin of this change in Chinese response dates back to the first millennium BC, during the Zhou dynasty, based on what Chinese philosophers call the Mandate from Heaven. This philosophy holds that Heaven gives wise and virtuous leaders a mandate to rule, and this mandate is removed only if the leaders become evil or corrupt. According to Taoist philosophy, Heaven expresses its disapproval of bad rule through natural disasters such as floods, plagues, or earthquakes.

On March 22, 1966, 17 years after the establishment of Communist rule and just prior to the social turmoil of the Cultural Revolution, the densely populated North China Plain, 200 miles (320 km) southwest of the capital city of Beijing was struck by the Xingtai earthquake of magnitude 7.2, causing 8000 deaths. This earthquake was followed the next year by the Hejian earthquake of magnitude 6.5, northeast of the Xingtai earthquake in the direction of Beijing.

Following this earthquake, Premier Zhou Enlai, possibly influenced by the Taoist Mandate from Heaven, issued the following charge to Chinese scientists:

> There have been numerous records of earthquake disasters
> preserved in ancient China, but the experiences are insufficient. It
> is hoped that you can summarize such experiences and will be able
> to solve this problem during this generation.

Premier Zhou's call for action may be compared with President John F. Kennedy's call a few years earlier to put a man on the moon by the end of the 1960s. Zhou had been impressed by the stories told him by survivors of the Xingtai earthquake, and he urged an earthquake prediction program "applying both indigenous and modern methods and relying on the broad masses of the people." China developed technical expertise, but also involved thousands of peasants who monitored water wells and observed the strange behavior of animals before an earthquake. This reflected Chairman Mao Zedong's belief that "humans will surely vanquish nature."

Mao did not trust the scientific establishment the Communists inherited from pre-revolutionary days, and in 1971 he created a new, independent government agency, the State Seismological Bureau (SSB), following the establishment of an Earthquake Work Leadership Group in 1969. The SSB reported directly to the Premier and had the status of an independent government ministry. Zhou's call for action led to a national commitment to earthquake research unmatched by any other country, including the United States. In 1981, the SSB was renamed the China Earthquake Administration (CEA), which contains well-equipped laboratories in every province in China, including, of course, the provinces of Shaanxi and Shanxi. The CEA employs thousands of workers, and its seismograph and geodetic networks cover the entire country. It is the largest earthquake-research organization on Earth.

The Chinese were not simply responding to the problem using brute force based on their large population. In 2008, then-Premier Wen Jiabao, a geologist by training, spoke of the need for "creative thinking and critical thinking" and "the integration of advanced civilization with traditional Chinese culture." In other words, use the scientific method, developed in the West, but use it in a uniquely Chinese way.

As a result, Chinese scientists of the CEA finally investigated the sources of the great earthquakes of 1303 and 1556, as well as other earthquakes that had struck China during its long history. In addition to their own investigations, Chinese earthquake geologists spent time at Western laboratories, including the Massachusetts Institute of Technology, where my colleagues Deng Qidong and Zhang Peizhen had studied, and the Institut de Physique du Globe de Paris, where I spent six months in 1993 and where another CEA colleague, Xu Xiwei, had worked with a French group that had done considerable research in western China.

Professor Xu invited me to visit the areas that had been damaged by the earthquakes of 1303 and 1556 and see what the CEA scientists had learned about those earthquakes. We traveled south through the major coal-producing city of Taiyuan to the city of Linfen, which lies within a triangle-shaped basin between the Lu Lian Shan (Lu Lian Mountains) on the west and the Taihang Shan on the east. The earthquake of 1303 accompanied rupture of the Huoshan fault on the northern margin of the Linfen basin, producing displacements as large as 25 feet (8 m). The fault scarp is still there, and we were able to walk along this fault and think about the catastrophic loss of life when the fault ruptured. An unusual feature was the length of the fault, only about 30 miles (45 km), much shorter than most faults that have produced earthquakes of magnitude 8.

We continued to the southwest and crossed the Yellow River near the mouth of the Wei River, and we soon came to the epicenter of the great Huaxian earthquake of 1556. This earthquake was caused by rupture of the Huashan fault, one of the faults at the base of the east–west-trending Qinling Shan to the south. As before, we were able to walk along the Huashan fault, which had been mapped in detail by the CEA. The rupture in 1556 was about 50 miles (80 km) long. Like the 1303 rupture, the break was shorter than expected for an earthquake of magnitude 8.

Although the culprit faults have been identified for both earthquakes, many questions remain unanswered. First, they are unusually short faults to produce earthquakes of magnitude 8. Second, neither of these faults, or any others in the loess hills of the Ordos region that have generated earthquakes, are on a plate boundary. They are within the Eurasian tectonic plate, and their long-term displacement rates are relatively slow. There are other active faults in the region that have not generated large earthquakes during China's long history, indicating that the next gigantic earthquake could break a fault that has not ruptured in China's recorded history. For this reason, the CEA has been unable to answer the late Premier Zhou Enlai's charge to "solve this problem during this generation."

In fact, no earthquakes of magnitude greater than 5 have struck the Ordos region in the past 100 years, and no catastrophic earthquakes have struck there in the past two and a half centuries. This observation clearly characterizes the Ordos earthquakes as an *earthquake cluster*. Clearly, these faults will rupture again, but the timing is even less clear than it is on the Enriquillo fault in Haiti: tomorrow or a century (or two or three centuries) from now.

The significance of these earthquakes is that they struck densely populated regions hundreds of years ago, similar to the heavily populated cities of the developing world today. Even though Chang'an was the cultural center of China's Middle Kingdom, most inhabitants lived in dwellings that were unable to resist the strong shaking of a major earthquake. The Yinchuan–Pingluo earthquake of magnitude 8 in the Ordos on January 3, 1739 ruptured a fault north of Xi'an that also offset the Great Wall of China. The population density was lower than that in the vicinity of the 1303 and 1556 earthquakes, and, accordingly, the loss of life was lower.

Zhou Enlai's call for action also proposed that Chinese scientists should learn to predict earthquakes, just as the programs established in Japan and the United States also had prediction as one of their goals (see Chapter 4).

By late 1974, none of the earthquake programs worldwide had predicted a major earthquake, but it appeared that China might actually be on the verge of an earthquake prediction (Wang, 2006). South of the city of Haicheng in southern Manchuria, the ground began to rise at an astonishing rate, and there were fluctuations in the Earth's magnetic field. Seismicity increased, and a long-range forecast predicted a moderate-size earthquake in that region in the next two years. Earthquake information was distributed to the public, and thousands of amateur observation posts were established to monitor various phenomena, including the strange behavior of animals. The forecast was modified in January 1975, for a moderate earthquake somewhere in southern Manchuria, northeast of Beijing.

Anomalous activity increased in the beginning of February 1975, including seismicity that caused alarm among the local people, including the citizens of Haicheng. More than 500 earthquakes were recorded early on February 4, and the provincial government issued a short-term earthquake alert, advising Haicheng's residents to move outdoors during the unusually warm evening of February 4. The large number of small earthquakes made this alert easy to enforce; it would have been more difficult to keep people indoors while the shaking was going on. At 7:36 p.m., the city of Haicheng was struck by an earthquake of magnitude 7.3, causing more than 90% of the buildings in parts of the city to collapse. Two thousand people lost their lives, but for a population of several million in the earthquake zone, the losses would have been much larger without the prediction. China had issued the world's first successful earthquake prediction!

While the Chinese were still celebrating their successful prediction, less than a year and a half later, on July 28, 1976, a larger earthquake of magnitude 7.8 destroyed the large industrial city of Tangshan, 180 miles southwest of Haicheng and closer to the capital city of Beijing (Figure 12.3). More than 240,000 people lost their lives, the worst natural disaster in modern times in China. Unlike Haicheng, there were no foreshocks, no warnings. China's earthquake

FIGURE 12.3 Destruction in 1976 of the library of the Hebei Mine Smelting Institute, City of Tangshan, due to earthquake. Source: photo by R. E. Wallace from USGS Photographic Library and *Earthquake Information Bulletin*, v. 15, no. 3.

experts (as well as the rest of the world) recognized that despite the Haicheng prediction, the problem of earthquake prediction remains unsolved. Haicheng, with its numerous foreshocks, was a special case.

The Haicheng earthquake was part of an *earthquake swarm*, in which the earthquake does not consist of a mainshock and aftershock sequence, like the Tangshan earthquake, but is an extended series of smaller earthquakes, in some cases not accompanied by a mainshock. The Haicheng sequence had its largest earthquake in the middle, not the beginning, but the Chinese still considered it as an ordinary earthquake. Other earthquake swarms lack a large earthquake, either at the beginning or within the sequence. The Tangshan earthquake, although it struck in the same region, was a typical mainshock–aftershock sequence.

In terms of the successful earthquake prediction, the Chinese clearly got lucky. Many similar earthquake swarms around the world did not have a large event in the middle of the swarm, as Haicheng (and, later, L'Aquila, Italy) did, but instead die down without a damaging earthquake. One member of a committee assembled by the Italians to advise the citizens of L'Aquila about the earthquake swarm affecting their city advised the people to go home and enjoy a glass of good Italian wine and not to worry. When a larger earthquake did strike, causing loss of life, the entire committee was charged and unjustly convicted of misinforming the public!

The Tangshan earthquake did lead to new insights into the earthquake distribution in the North China Plain. Professor An Yin of UCLA and his Chinese colleagues observed that the Xingtai, Hejian, and Tangshan earthquakes lined up in a northeast–southwest direction and included the 1830 Cixian earthquake of magnitude 7.5, southwest of the Xingtai earthquake (Figure 12.1). The Tangshan earthquake itself was part of an earthquake sequence that trends northeast–southwest. An Yin and his research colleagues called this line-up of earthquakes the Tangshan–Hejian–Cixian fault, characterized by right-lateral strike-slip. This fault extends through the megacity of Tianjin, with a population of 22 million, but Tianjin and its seismic gap 100 miles (160 km) long has not had a large earthquake on this fault in 8400 years. Slip rates on faults in the North China Plain are slow, and the new understanding of earthquake potential suggests that the next earthquake could strike in the next 2000–3000 years if the gap were filled by a single earthquake. (Alternatively, the gap could be filled by more than one earthquake or by a creeping fault.) In any event, I list the city of Tianjin as an earthquake time bomb.

What about the Mandate from Heaven? Demonstrations in Tiananmen Square the preceding March had been precipitated by people laying wreaths in honor of the recently deceased Premier Zhou Enlai and demonstrating against the excesses of the Cultural Revolution and the seizing of power by an ideological group of extremists called the Gang of Four, including the wife of Chairman Mao Zedong. The demonstrations were brutally put down by the military, and one of the more practical leaders, Deng Xiaoping, was exiled.

Shortly after the Tangshan earthquake, Mao Zedong died and was replaced by Hua Guofeng. This set up a confrontation between the Gang of Four and Hua, which resulted in the arrest of the Gang of Four and the return of Deng Xiaoping to power in 1977. Deng introduced a policy of modernization that has made China a superpower, with an economy second only to that of the United States and with living standards for many Chinese citizens much higher than they have ever had in their long history. One could argue that the Mandate from Heaven has been carried out!

Still, China's earthquakes persist. On May 12, 2008, Sichuan Province in southwest China was struck by the Wenchuan earthquake of magnitude 7.9, with a loss of life of more than 87,000 people. Like Tangshan, it had not been predicted, and the source fault that generated the earthquake had not been identified as a high-risk fault.

The large losses in 1303 and 1556 were due to the large population in two of the world's first megacities. They were not the only megacities that developed prior to modern times. Rome, Italy, and Alexandria, Egypt, were megacities 2000 years ago, but fortunately for their inhabitants they were not visited by a great earthquake while they had large populations. Alexandria is not in a region where one would expect great earthquakes, although it was heavily damaged by a tsunami in the fourth century AD. Rome is in earthquake country, but fortunately for the Romans their capital city has not been struck by a major earthquake since the time Rome was first heavily populated.

The Chinese were not so lucky.

Other time bombs, including cities
that are not well prepared

13 Age of Enlightenment and the 1755 Lisbon earthquake

The seventeenth and eighteenth centuries, following the Protestant Reformation and the Catholic Counter-Reformation, were a time of great intellectual ferment in Western Europe. Philosophers challenged ideas based on tradition and faith, and they urged their colleagues to show skepticism about long-term religious dogma and the concept of absolute monarchy. Its early practitioners included the philosophers Baruch Spinoza and John Locke in the late seventeenth century, followed by Voltaire, Jean-Jacques Rousseau, and Thomas Paine in the eighteenth century.

The Age of Enlightenment was accompanied by a Scientific Revolution, including a physicist, Isaac Newton, a geologist, James Hutton, and a mathematician, Leonhard Euler. Eventually, the Age of Enlightenment spread to the American colonies, led by Benjamin Franklin and Thomas Jefferson, both of whom were scientists as well as political leaders. The American Declaration of Independence, the US Constitution, and the Bill of Rights are direct products of the Age of Enlightenment as Americans sought to establish an ideal democracy in the New World.

For Immanuel Kant, in an essay in 1784, enlightenment marked the liberation of human consciousness from a state of ignorance to a state of reason. These views swept across Europe, and, among other results, led to a scientific treatment of disease. As described elsewhere (*Population Explosion and Increased Earthquake Risk to Megacities*), Enlightenment and the Scientific Revolution were accompanied by an improvement of medical care and health, which itself led to people living longer and to an increase in population that continues today.

THE LISBON EARTHQUAKE

At 9:30 a.m. on November 1, 1755, All Saints Day, the views of Enlightenment philosophers were subjected to a major crisis. As described by Fonseca (2004), Lisbon, the capital city of Portugal, a maritime superpower with colonies in Africa, Asia, and South America, was destroyed in a few minutes by a massive earthquake (Figure 13.1). The earthquake was felt in many parts of Western Europe and North Africa (Figure 13.2). Western society was shocked that one of the most beautiful and prosperous cities in Europe could suffer such a fate. About 40,000 people, one-fifth of Lisbon's population, lost their lives, and another 10,000 died in Morocco to the south. A subsequent report gave the losses in Lisbon as nearly 100,000.

People rushed to the Lisbon harbor, where the sea had drained away, revealing debris, including lost cargo and previously wrecked

FIGURE 13.1 The Lisbon earthquake of 1755, illustrating the collapse of buildings and the great tsunami that swept through the harbor. Source: Bettmann Archives through Corbis Images.

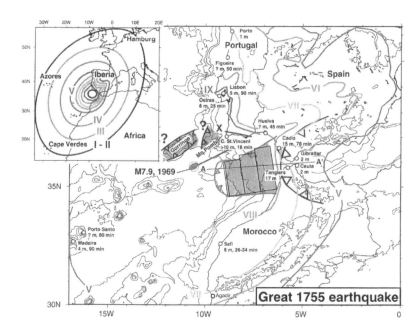

FIGURE 13.2 Setting of the great 1755 Lisbon earthquake. There appears to be a subduction zone between Spain and Morocco, shown as a sharply curved line with triangles. Roman numerals show Modified Mercalli Intensity (MMI), including the inset map. Also shown are the maximum height of the 1755 tsunami, in meters (1 meter = about 3 feet) and giving the time between the earthquake and the arrival of the tsunami. Note location of earthquake of magnitude 7.9 in 1969. Source: from Yeats (2012, Fig. 6.6) and Gutscher *et al.* (2006) with permission from Elsevier.

ships. A short time after the earthquake, the sea rushed back into the harbor in a massive tsunami, drowning many people. People on horseback were barely able to gallop to safety ahead of the waves. The tsunami caused great damage in Morocco, Algeria, and the Algarve coast of southern Portugal, and it was recorded in the West Indies, Brazil, and Scotland. After the tsunami receded, fires raged in Lisbon for the next five days.

Religious authorities suggested that the earthquake and tsunami were the result of divine judgment, but it was soon pointed out that Lisbon's red-light district, called the Alfama, had suffered only minor damage. Voltaire used the earthquake in his most famous written work, *Candide*, published three years after the earthquake, to argue that this was not the best of all possible worlds, supervised by a benevolent God. Jean-Jacques Rousseau pointed out that the earthquake, because of its massive destruction of the heavily populated city of Lisbon, showed the peril of living in crowded cities, arguing for a more naturalistic way of life. (This book also points out the danger of living in megacities, using some of the arguments first made more than two and a half centuries ago by Rousseau.) Immanuel Kant wrote three essays on the Lisbon earthquake, collecting all the information he could about the earthquake and coming up with a theory for the origin of earthquakes.

The magnitude of the earthquake may have been as large as 8.6–8.8. The tsunami indicates that the source of the earthquake was offshore, possibly on a subduction zone west of the Strait of Gibraltar in which the Atlantic oceanic crust is being subducted eastward beneath the Strait of Gibraltar, the Iberian Peninsula, and Morocco (Figure 13.2). Other earthquakes have been reported in this region, including an earthquake of magnitude 7.9 in 1969. However, despite considerable work in the past two decades and despite the essays written by Immanuel Kant, the source of the 1755 earthquake is still poorly understood.

There might have been two earthquakes, one off Portugal and one off Morocco, as suggested by the bowing out of lines of equal seismic intensity in Morocco (Figure 13.2). Shaking from two separate earthquakes might not have been confirmed at the time because of poor relations and poor communications between Christians of Western Europe and Muslims of North Africa.

The tectonic setting is complicated. The Strait of Gibraltar is at the western end of a major tectonic plate boundary between Europe and Asia to the north and Africa, Arabia, and India to the south. Eurasia is separated geologically from Africa, Arabia and India (Gondwanaland) by the Tethyan ocean, which has been tectonically closed up between India and the Himalaya but still exists as the

Mediterranean Sea south of Europe. North-dipping subduction zones underlie Italy and Greece and generate active volcanoes, including Vesuvius and Etna in Italy and Santorini in the Aegean Sea. The subduction zone south of Greece generated an earthquake of magnitude greater than 8 in AD 365, generating a tsunami that heavily damaged the city of Alexandria in Egypt. Several tsunamis have struck the coast of Palestine and Lebanon, some of them possibly from subduction zone sources in the Mediterranean to the west.

The east–west tectonic boundary between Europe and Africa is not the source of the Lisbon earthquake. The east–west boundary joins a possible north–south boundary between the Mediterranean and Atlantic Ocean in a feature called the Gibraltar subduction zone. This may have been the source of the 1755 earthquake, although the length of the Gibraltar subduction zone should have limited the size of the earthquake to a magnitude no larger than 8. The reason for the large magnitude is unexplained, although more than one earthquake might be the answer.

We know a little bit more today about the tectonic setting of the earthquake than Kant did in 1755, but not much.

RESPONSE TO THE EARTHQUAKE

This brings us to Sebastião José de Carvalho e Melo, the Head of the Government of Portugal, knighted by royal decree by King Joseph I four years after the earthquake as the first Marquês de Pombal. At the time of the earthquake, Melo was Secretary of State of the Kingdom of Portugal and the Algarves, a rank equivalent to Prime Minister. In 1738, Melo had been appointed Ambassador to Great Britain, which permitted him to view in admiration the economic success of the English. Upon the accession of Joseph I to the Portuguese throne, Melo was appointed Foreign Minister, and in 1755 he was elevated to an office equivalent to Prime Minister, which he held at the time of the earthquake.

Fortunately, Melo survived the earthquake, and he took it upon himself to rebuild the city. He issued a statement that became famous:

"What now? We bury the dead and heal the living." He mobilized troops, obtained shelters, and arranged for help for survivors of the earthquake. Almost immediately, he began plans to rebuild the city.

First, he designed a questionnaire to be sent to parishes throughout Portugal, asking people to report on whether water levels in wells had risen or fallen during the earthquake, and whether respondents had noted strange behavior of animals, especially dogs, and on the destruction of buildings where people lived. The results of this survey, the first of its kind, have allowed modern seismologists to understand in greater detail the 1755 earthquake. Such surveys are still conducted today. The USGS website has a questionnaire called "Did You Feel It?" for people to fill out online after they experience an earthquake (see Chapter 2).

Lisbon did not experience an epidemic after the earthquake, and so Melo could set about reconstruction of the city. The central part of Lisbon was rebuilt to withstand future earthquakes. Architectural models were constructed and tested by having soldiers march around the models, a forerunner to modern shake tables. As a result, the Pombelline Downtown (*Pombelline Baixa*), a major tourist attraction today, became the world's first district reinforced against earthquakes. One of the structures was called the Pombelline Cage, a symmetrical wood-lattice structure designed to distribute earthquake forces in buildings.

The reforms instituted by Melo, including the earthquake questionnaire and the construction of seismic-resistant buildings in central Lisbon, were consistent with the scientific reforms accompanying the Age of Enlightenment, and Melo is commonly considered a true product of the Enlightenment and Scientific Revolution. However, as prime minister, Melo ruled with a heavy hand, acting as a virtual dictator during much of his time in office. The laws he passed made him many enemies, particularly among the Jesuits and the nobility. Even so, the attention to seismic strengthening and to a better understanding of earthquakes would not have been possible without the other changes taking place throughout Europe and the North American colonies as a result of the Age of Enlightenment.

14 Jerusalem: earthquakes in the Holy Land

INTRODUCTION

The preceding chapters on China and the Age of Enlightenment focused on the destruction of a major city, Xi'an, with millions of inhabitants four and a half centuries ago and the destruction of another major city, Lisbon, in 1755. This chapter steps farther back in time to discuss earthquakes in Israel/Palestine, including some that took place more than 3000 years ago. The population of Jerusalem is estimated as 25,000 in the days of King Solomon in the tenth century BC, 60,000 in AD 70, 30,000 in AD 1130, and fewer than 9000 in AD 1800. The population began to grow in the nineteenth century, but was still only 164,000 in 1946, two years before the establishment of the State of Israel. This means that earthquake losses throughout history in Jerusalem have not been related to a population explosion. The rapid increase in population did not start until after World War II, and only recently has the population of Jerusalem risen to the level of hundreds of thousands. Jerusalem's date with its next earthquake, when many people will be killed, is still in the future.

I include the city of Jerusalem as a time bomb because it is sacred to three of the world's great religions, giving it a prominence far greater than its population of three-quarters of a million would otherwise justify. About two-thirds of its population are Jewish, and nearly all the rest are Arabs, mainly Muslim, but also Christian. The higher birth rate among Arab families and the departure of secular Jews from the city result in a growing percentage of Arabs in Jerusalem's population, although this trend could reverse itself because more ultra-Orthodox Jewish families, also with high birth rates, are moving into the city.

A major social problem is that both Israel and the Palestinian Authority claim Jerusalem as their capital, and at present, despite peace talks over the years, there is no compromise to share the city that would lead to an Arab–Israeli peace. A wall has been built between Arab communities and housing developments for Jewish settlers within the West Bank. Commentators worry about the possibility of a new *intifada*, or uprising, if no peace agreement is reached. Renewed violence is assured if extremist Jewish parties take control of the government and attempt to annex the West Bank to Israel.

But Jerusalem has another critical problem that is almost never mentioned in the media: earthquakes. The city is only 35 miles (60 kilometers) west of the Dead Sea fault, which forms the boundary between the Arabian and African tectonic plates and is located in the valley of the Jordan River (Figure 14.1). The Arabian plate is being driven northward past the African plate, including Palestine and the eastern Mediterranean region, against the high mountains between Turkey and Iran, including the mountains of the Caucasus region. The motion of the plates leading to the uplift of these mountains is slower than a fifth of an inch per year (4–5 mm yr^{-1}) – slower than the rate your fingernails grow.

This sounds like a rate so slow that it can be ignored, but the history of Palestine argues otherwise because Israel/Palestine, including what is now modern Israel and the West Bank, has been wracked by destructive earthquakes several times during the past 3000 years. To this history is added the evidence from archaeology, which extends the record back even farther in time, including the history of Jericho, one of the world's oldest cities. Some of the earthquakes have been described in the Bible.

LACK OF CONFIRMATION OF BIBLICAL RECORD BY ARCHAEOLOGY

Before using biblical evidence for earthquakes, it is necessary to deal with the lack of written or archaeological confirmation of biblical events earlier than about 3000 years ago. Contemporary written

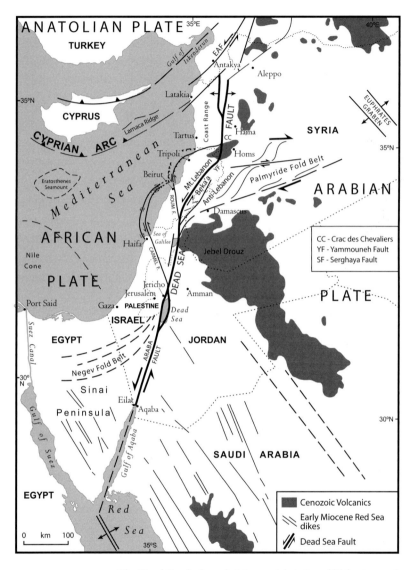

FIGURE 14.1 The Dead Sea fault and cities at risk in Israel/Palestine and Levant (Syria and Lebanon). CC, Crac des Chevaliers; Serghaya fault (southeast of Beka'a Valley); YF, Yammouneh fault. At the northern end of the map, faults turn to the northeast as the East Anatolian fault (EAF). Source: modified from Yeats (2012, Fig. 7.15) and Searle *et al.* (2010).

records are available for this earlier period from Egypt to the west and from Mesopotamia (present-day Iraq) to the east, but not from Palestine, even though the climate at that time was suitable for societies to grow and keep records, as they did elsewhere. The lack of archaeological confirmation of the stories of Noah, Abraham, Moses, Joshua, and even King David and King Solomon has produced a major controversy among some of the world's leading archaeologists within Israel and elsewhere. The debate about the very existence of these Old Testament figures is political as well as scientific, because the biblical accounts of David and Solomon and their predecessors are essential to Israel's claim to Zion, the "Promised Land," promised to the Israelites by Jehovah.

The early books of the Old Testament are based on oral traditions passed down from generation to generation, and a locally written history of Israel and Judah did not become available until after the time of David and Solomon. For this reason, scientists have turned to the rich archaeological record, focusing on the reigns of David and Solomon, because as rulers of a regional empire they should have constructed large buildings, including a royal palace, the ruins of which could be uncovered. Controversies arise with excavated buildings that were presumed to have been constructed at the time of David and Solomon, between 1010 BC, when David became king, to around 930 BC, the time of the death of his son, Solomon. Additional questions have been raised about whether ruins currently under excavation, attributed to the time of David and Solomon, are younger, possibly as young as the time of the conquest of Israel and Judah by the Assyrians.

This controversy is reviewed by Robert Draper in the December 2010 *National Geographic Magazine*. For many years, the words of the Old Testament were used as evidence to corroborate archaeological interpretations. However, this may be an example of circular reasoning, and the present debate focuses on evidence other than the written biblical record, which is largely based on oral traditions written down long afterwards.

Archaeologist Eilat Mazar announced in 2005 that a low stone wall adjacent to a retaining wall in the Kidron Valley close to modern Jerusalem is part of the ruins of King David's palace as described in the second book of Samuel. However, archaeologist David Ilan of Hebrew Union College doubts that Mazar has demonstrated conclusively that this was indeed David's palace. The archaeological ties to David and Solomon have also been questioned by archaeologist Israel Finkelstein of Tel Aviv University, who concludes that these buildings were built one or two centuries later.

But in the Elah Valley, southwest of Jerusalem, the place where David was reported to have killed the Philistine giant Goliath, Yosef Garfinkel of Hebrew University claims to have found a corner of a building constructed at the time of David, with its age confirmed by radiocarbon dates of olive pits from the site. Others have been excavating ancient copper mines east of the Jordan River that appear to have supported a major empire. In Garfinkel's view, these discoveries vindicate the earlier work using both archaeological evidence and biblical scholarship.

But the opposition remains unconvinced, and the debate rages on. The accounts in the Bible of earthquakes during this early period are not true contemporary accounts but should be considered as based on verbal stories written down centuries later. This applies to the biblical story of the destruction of Sodom and Gomorrah and of Jericho, as described in a book by David Neev of the Geological Survey of Israel and the late K. O. Emery of Woods Hole Oceanographic Institution, published in 1995 by Oxford University Press.

For more recent earthquakes in the region, see Ambraseys (2009).

DESTRUCTION OF SODOM AND GOMORRAH

These cities were probably both within the Dead Sea Basin, itself formed by strands of the Dead Sea fault (Figure 14.2). The Dead Sea itself is the lowest place at the surface of the Earth not covered by seawater. Sediments deposited in the Dead Sea Basin, with evidence for contemporary earthquakes over a period of 70,000 years, have been

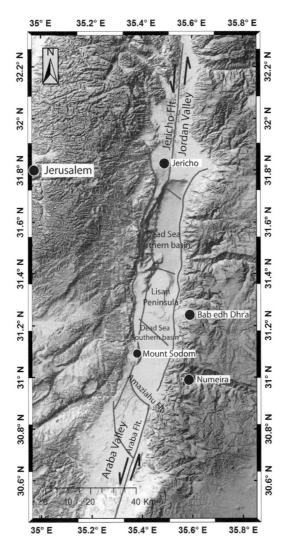

FIGURE 14.2 Computer-generated topographic map of the Dead Sea region, including topography beneath the Dead Sea. The South Basin is now filled with sediments; North Basin is still a body of water. Numeira and Bab edh Dhr'a are archaeological sites that experienced a large earthquake at about the same time as the biblical account of the destruction of Sodom and Gomorrah. The proximity of Jerusalem to the Dead Sea fault zone indicates the hazard of that city to a plate-boundary earthquake. Source: map from Hall (1994).

mapped and dated by Shmuel Marco and his colleagues at Tel Aviv University, where the sediments are called the Lisan Formation, after the Lisan Peninsula that divides the Dead Sea into a north basin and south basin.

South of the Dead Sea, the local name for the Dead Sea fault is the Wadi Araba (Arava) fault, which extends southward along the Wadi Araba 100 miles to the Gulf of Aqaba (Figure 14.1), which was the site of an earthquake in Egyptian territory of magnitude 7.3 in 1995.

North of the modern city of Aqaba in Jordan and its twin city, Eilat, in Israel, the Wadi Araba fault was the source of an earthquake in AD 1068 of magnitude 7 or larger, which destroyed a city at the location of modern Eilat and damaged the city of Medina to the south in present-day Saudi Arabia. There may have been two earthquakes that year, with the other one heavily damaging the city of Ramla, southwest of Jerusalem. An earthquake in AD 1212 of magnitude 6.5–7, with its epicenter in the Gulf of Aqaba, damaged settlements in the Wadi Araba and caused the collapse of the church and monastery of St. Catherine in the Sinai Peninsula of Egypt. An earthquake in AD 1293 of magnitude 6.5–7 damaged a castle just above the Dead Sea and produced destruction as far west as Gaza on the Mediterranean coast and Ramla, southwest of Jerusalem. Finally, an earthquake in AD 1458 of about the same size in the same region caused offset along the Wadi Araba fault of a water tank that was part of the Tilah Castle, built during Roman or early Byzantine time. This earthquake destroyed minarets in Ramla and Jerusalem, and a dome near the Church of the Holy Sepulchre was destroyed. Evidence for all four of these earthquakes is recorded in paleoseismic trenches in sediments of the Ze'elim Formation, overlying the Lisan Formation in the Dead Sea Basin, sediments that have been deformed by earthquakes.

This brings us to Sodom and Gomorrah and its destruction, as recorded in the Book of Genesis. According to Genesis, Abraham and his family, including his nephew, Lot, and their herds, reached the "cities of the plain" in what is now believed to be the southern Dead

Sea Basin. The region was in the midst of a famine, and Lot pitched his tents near the city of Sodom. The "cities of the plain" were engaged in a war with invaders from Syria, and the kings of Sodom and Gomorrah, along with Lot, were captured and taken to the north. Abraham pursued the invaders and defeated them, returning the prisoners to their homes in the Dead Sea region, where they offered hospitality to Lot and his family. However, residents of Sodom, where Lot's family was to spend the night, demanded "homosexual access" to their guests, who escaped with the help of angels, and fled to the east.

The next day, Lot and his family left for another "city of the plain" called Zoar. At that time:

> the Lord rained upon Sodom and upon Gomorrah brimstone and fire from the Lord out of heaven; and he overthrew those cities, and all the plain, and all the inhabitants of the cities, and that which grew upon the ground ... and lo, the smoke of the country went up as the smoke of a furnace.
>
> *(Genesis 19:24, 25, 28, King James Bible)*

Four of the five "cities of the plain" were destroyed, and only Zoar was spared. This is one of the most extreme examples of destruction in the Bible, and it probably describes an earthquake (although my late colleague, Nicholas Ambraseys, was not convinced). The outburst of smoke and sulfurous fire has been attributed to the igniting of asphalt, which is known to be present around the Dead Sea, but an alternate theory I favor is that the "smoke" was due to great clouds of dust that commonly accompany earthquakes in the desert, with a recent example being a large earthquake southeast of San Diego in the desert of Baja California, Mexico, in 2010.

The timing of the destruction of Sodom and Gomorrah is estimated as 2300–1700 BC, during the archaeological age known as Early Bronze Age III. The Bab edh Dhr'a and Numeira archaeological sites east of the southern Dead Sea Basin (Figure 14.2) were struck by two large earthquakes within a period of 50 years about that time. One of these might have been the biblical earthquake that destroyed Sodom and Gomorrah. A salt-cored mountain on the west edge of the southern Dead Sea Basin is named Mount Sodom (Figure 14.2, alternatively Mount Sedom). The ages and environment are consistent with this interpretation, but without a written contemporary record with names and places, the biblical story cannot be confirmed. Bab edh Dhr'a and Numeira might have been the ancient cities of Sodom and Gomorrah, or they might not.

JOSHUA AND THE WALLS OF JERICHO

Jericho, part of the Palestinian West Bank, is one of the oldest cities on Earth, possibly as old as 10,000 years. An archaeological mound, called Tell Jericho or Tell el-Sultan, is close to the Jericho fault, part of the Dead Sea fault system (Figure 14.2). The location of the city may be due to a fresh-water spring called the Spring of Elisha, so named because it was said to have been "purified" by the prophet Elisha (2 Kings 2:19–22). The city had been heavily damaged by earthquakes several times in its history, including destructive earthquakes 7500 and 7000 years ago, based on archaeological excavations. The inhabitants kept coming back, though, because of the Spring of Elisha, their source of water in one of the driest places on Earth.

Did the walls of Jericho collapse during an earthquake when the city was attacked by the Israelites under Joshua, as described in Joshua 3:13–16? A more fundamental question is whether the invasion of Canaan by Joshua and the Israelites was a historical event, part of the question of the existence of Moses and Aaron and the Israelites fleeing the Pharaoh of Egypt. Jericho had been conquered by the Egyptians around 3560 years ago and resettled by Late Bronze Age people around 3400 years ago. The city was destroyed and abandoned after attacks by Asiatic nomads around 3300 years ago. These nomads could have been rebellious slaves escaping Egypt during the reign of Ramses II. This was possibly the origin of the oral tradition of the invasion of Canaan by the Israelites led by Joshua. However, there is no archaeological evidence supporting the conquest of Jericho as written in the Bible. Nor is there archaeological evidence for the destruction of the city wall or the subsequent burning of the city (Joshua 6:20–24) at that time.

Joshua 3:13–16 reports that Joshua and his army crossed the Jordan River near Jericho, 20 miles south of the biblical city of Adam (modern Damieh), where the flow of the river had stopped, allowing the Israelites to cross. The Jordan River also temporarily dried up as the result of a landslide at Damieh in AD 1267 and again at the time of an earthquake of magnitude 6.3 near Jericho on July 11, 1927. That earthquake was accompanied by a landslide at Damieh that stopped the flow of the Jordan River for 22 hours.

There is archaeological evidence for the burning of Jericho at the end of the Early Bronze and Middle Bronze ages, but not at the end of the Late Bronze age, the generally accepted age of the Exodus of the Israelites. So the collapse of the walls of Jericho at the hands of Joshua and the Israelites is not confirmed by archaeology.

EARTHQUAKE OF 759 BC

The prophet Zechariah issued the world's first earthquake forecast when he wrote (Zechariah 14:4–5, King James Bible):

> And his feet shall stand in that day upon the mount of Olives, which is before Jerusalem on the east, and the mount of Olives shall cleave in the midst thereof toward the east and toward the west; and there shall be a very great valley; and half of the mountain shall remove toward the north and half of it toward the south. And ye shall flee to the valley of the mountains; for the valley of the mountains shall reach unto Azal; yea, ye shall flee, like as ye fled from before the earthquake in the days of Uzziah king of Judah.

An earthquake of magnitude 7.3 struck probably on October 11, 759 BC, heavily damaging Solomon's temple. It was reported in other books of the Old Testament (Amos, Isaiah), as well as the Talmudic literature. The exact date is based on a solar eclipse on June 15, 763 BC. 2 Chronicles reports that the earthquake occurred on the eve of the Tabernacle holiday, Tishrei 14, 3003, which corresponds to

October 11, 759 BC. The epicenter may have been near biblical Hazor, north of the Sea of Galilee (SG, Figure 14.1), where archaeological evidence shows tilting of northern and western walls of excavations of that age. However, archaeological excavations and places in biblical Judah and Israel named in the literature are difficult to assign to the 759 BC earthquake, in part because of an invasion of Palestine by the Egyptian pharaoh Sheshonk the First. Amos reports suggestions of a tsunami (seismic sea wave) in the Sea of Galilee, and the city of Kinnereth on the north shore of the Sea of Galilee was destroyed.

However, the details of Zechariah's description do not hold up to modern geological mapping by the Geological Survey of Israel, which shows a major landslide but no large strike-slip fault on the Mount of Olives east of Jerusalem. The Atlas of Israel does show a fault southeast of Jerusalem that passes through the Mount of Olives, but this fault trends east–northeast, not north–south, and it is not clear that it is an active fault.

EARTHQUAKES IN 31 BC AND AD 33

Zechariah's earthquake forecast was presumably fulfilled by an earthquake in 31 BC as reported in AD 93 by the Roman-Jewish historian Josephus, resulting in the deaths of 30,000 men. But modern scholars believe that the severity of the 31 BC earthquake, including the death toll, was greatly exaggerated.

The Gospel according to Matthew describes two earthquakes in AD 33, the first during the Crucifixion and the second after the Resurrection, allowing women to enter the tomb of Christ and verify the absence of his body. Among the four Gospel writers, only Matthew mentions earthquakes, and verification by contemporary writers or by archaeological evidence is inconclusive. There is no written description of an earthquake affecting Jerusalem, although some suggest that the earthquake caused the opening of a rock beneath the Chapel of the Exaltation of the Cross. There are east–west-trending fractures in limestone at this location, but no independent evidence for surface faulting.

EARTHQUAKES OF AD 363

The late fourth century was a time of great earthquake activity in the eastern Mediterranean, called the Great Byzantine Paroxysm. Two earthquakes, six hours apart, struck first northern Palestine and adjacent Syria and then southern Palestine and sites east of the Dead Sea. These earthquakes destroyed 22 towns and caused great loss of life, although it has not been possible to distinguish which earthquake destroyed which towns. Heavy damage was reported in towns of northern Palestine, including Haifa, Jerusalem and its suburbs, and areas close to the Dead Sea, including Petra in modern Jordan and the ancient city of Zoar. The earthquake caused heavy damage as far north as Antioch. The earthquakes brought to a halt the restoration of the Temple at Jerusalem that had been authorized by the Roman Emperor Julian.

EARTHQUAKE OF AD 746–749

This very large earthquake (possibly two or three earthquakes within a few years), with magnitude estimated as 7.3, damaged 600 settlements in Palestine and destroyed the Hisham Palace in Jericho. The Aksa mosque in Jerusalem was damaged. An archaeological site at the Herodian city of Tiberias on the shore of the Sea of Galilee contains the remains of a Roman stadium and Byzantine and early Arab (Umayyad) structures, all of which were offset by the Dead Sea fault. Younger Arab (Abbasid) structures were undisturbed, confirming correlation of faulting with the AD 749 earthquake. Damage from this earthquake extended from Jericho in the south to Ba'albek in the Beka'a Valley of Lebanon to the north. Deformation related to the earthquake was found in a paleoseismic trench north of Aqaba, excavated by Yann Klinger of the Institut de Physique du Globe de Paris.

YOUNGER EARTHQUAKES IN THE NORTH

An earthquake of magnitude 7.5 in 1202 produced major damage in northern Palestine, including Tiberias, Acre, and Nablus, and in

Lebanon, including Ba'albek and Tripoli. This earthquake was part of the history of the Crusades in that a Crusader castle, one of the largest in the Latin kingdom of Jerusalem, was under construction within the Dead Sea fault zone, close to a ford of the Jordan River called Vadum Iacob. Construction began in 1178, but the partially completed castle was besieged and destroyed in 1187 by the Arab general Saladin. Fifteen years later, the northern and southern defense walls of the castle were displaced several feet (2.1 m) along the Dead Sea fault. A mosque built within the walls of the ruined Crusader castle was offset only about two feet (60 cm). The mosque was offset by a younger earthquake, probably in 1759.

Earthquakes characterized the Dead Sea fault north of Palestine, including Lebanon, western Syria, and southernmost Turkey. Antakya (Antioch) in southern Turkey was heavily damaged by earthquakes in AD 115 and 526. Aleppo, the largest city in modern Syria (Figure 14.1), was destroyed by an earthquake in 1138 that was reported to have killed 230,000 people, a number that is too high because the population of Aleppo at that time was only a few tens of thousands. The Dead Sea fault branches off to the northeast as the East Anatolian fault in southern Turkey, which has had major earthquakes in the past, including a series of earthquakes in the nineteenth century.

EARTHQUAKE HAZARDS IN THE HOLY LAND

The earthquakes described above, both those during biblical times and afterwards, indicate that all parts of the Dead Sea fault must be considered as seismically hazardous. Other earthquakes struck the region, but they are less well known. Scientists led by Shmuel Marco of Tel Aviv University and Amotz Agnon of Hebrew University, studying lake deposits of the Dead Sea that have been subjected to strong shaking, have worked out an earthquake history for the last 70,000 years, the longest earthquake record on Earth. Dates of earthquakes from lake deposits left over the past 2000 years can be correlated with historical earthquakes.

In addition to the Dead Sea plate-boundary fault itself, two additional faults branch off northwestward from the Dead Sea fault to the Mediterranean Sea (Figure 14.1). South of the Sea of Galilee, the Carmel fault extends from the Dead Sea fault and reaches the sea at Haifa, Israel's main port. The Roum fault to the north branches off north of the Sea of Galilee and reaches the sea at Beirut, capital city of Lebanon. Surface rupture on the Roum fault has been documented; it was the source of an earthquake of magnitude 7.1 in 1837. Offshore geophysical surveys show evidence for faults parallel to the coast between Haifa and Tripoli in Lebanon. One of these faults is the Mt. Lebanon thrust, which was the source of an earthquake of magnitude 7.5 in AD 551 that did great damage to Beirut. One must conclude that the area in harm's way includes most and probably all of the Holy Land. In addition to Jerusalem and the region south to the Gulf of Aqaba, the cities in danger include Ramallah, Hebron, and Nablus in the West Bank, Amman, the capital of Jordan, Beirut, the capital of Lebanon, Damascus, the capital of Syria, Aleppo, Syria's largest city, and Antakya (Antioch) in southern Turkey.

I have heard no discussion of earthquake-resistant construction in the region governed by the Palestinian Authority, and the heavy losses from the 1999 Izmit earthquake in Turkey because of poor construction practices are not encouraging. Israel, being a more-developed country with world-class earthquake scientists and engineers, has building codes that are stronger than those elsewhere in the region and are rigidly enforced. I do not know how this translates to the West Bank and the Gaza Strip. If damage and loss of life from the next inevitable earthquake are concentrated in the Arab part of Palestine, what effect would this have on Middle East politics?

In addition, the coast of the Gaza Strip, Israel, Lebanon, and Syria are at risk from tsunamis, or seismic sea waves. A tsunami in Gaza, the largest city in the Palestinian Territories, with a population of 450,000, could be catastrophic. Tsunamis have been attributed to the AD 551 earthquake on the Mt. Lebanon thrust. Tsunamis from the 1202 and 1759 earthquakes on the Dead Sea fault may have been triggered by submarine landslides caused by strong shaking. Other tsunamis were generated on subduction zones farther west, off Cyprus, Crete, and southern Italy. Does this coastal region need to learn the hard way, by losses of life through drowning, that it faces a tsunami hazard in addition to an earthquake hazard?

15 Istanbul: responding to an official earthquake warning

HISTORICAL BACKGROUND

There has been a settlement on the Bosphorus waterway connecting the Mediterranean Sea and Black Sea for more than 2600 years. All ships leaving Russia, the newly independent countries of the Caucasus, the Ukraine, Romania, and Bulgaria must pass through the Bosphorus Strait past the city of Istanbul to reach the Aegean Sea and the markets of the world. The Bosphorus is the boundary between Europe on the west and Asia on the east, although both sides are within the national borders of Turkey.

The strategic significance of the Bosphorus was not lost on the Greek sailors of the city-state of Megara, near Athens, when they founded Chalcedon on the Asian side of the Bosphorus in 685 BC. The Greeks soon recognized that the European west side had the better prospects for a city and port, and so 28 years later, the Megarans, led by King Byzas, founded a second town they called Byzantion, after their king, on the European side of the Bosphorus. The strategic location of the city led to further Greek exploration and colonization around the shores of the Black Sea. With the advent of the Romans, the town became known by its Latin name, Byzantium. It was part of the Roman province of Bithynia, on the south shores of the Black Sea.

In the days of the Roman Empire, Byzantium increased its importance, particularly after the Emperor Diocletian divided the Empire in AD 285 into western and eastern domains. During this time, the new Christian religion was growing in importance, and in AD 330, Christianity was adopted by Emperor Constantine I as the state religion of the Roman Empire. The city of Byzantium was renamed Constantinople after the emperor, a name it would keep for the next 1600 years until the name was officially changed to Istanbul. (However, the city had been unofficially referred to as Istanbul by Muslims as long ago as the tenth century.) When the western half of the empire, including Rome, was overrun, the remains of the Roman Empire became known as the Byzantine Empire. The main language of the Roman Empire had been Latin, but after the division of the empire and the collapse of its western half, the language and culture of the region reverted to Greek.

The truncated empire that was ruled from Constantinople was in the path of the Arab Muslim invasions in the seventh century, although the city was subsequently captured not by the Arabs but by members of the Fourth Crusade, diverted from an original plan to re-take Jerusalem from the Arabs. During this time, Constantinople was the capital of one of several Christian states in the region. Although the Byzantine Empire was later re-established, a more serious threat arrived with the Ottomans, also Muslim, who took over the remains of the Byzantine Empire. In 1453, the Ottomans under Mehmet II finally captured Constantinople, which then became the capital of the Ottoman Empire. The Ottomans under Sultan Sülayman the Magnificent launched a major offensive against Christendom, reaching a high-water mark with an unsuccessful attempt to capture Vienna.

In the nineteenth century, the Ottoman empire went into decline, and after World War I, in which the Ottomans backed the losing side, the sultanate was abolished. A republic was established in 1923 under the leadership of a young, charismatic leader, Mustafa Kemal, called Atatürk. Constantinople was officially renamed Istanbul in 1930, although the Greeks still refer to the city as Constantinopolis.

I spent time in Istanbul in the 1970s and found similarities to San Francisco. Both are among the world's most vibrant cities. The Bosphorus was crossed by a major bridge that reminded me of San Francisco's Golden Gate. It is now crossed by two bridges and a tunnel. Its many cultural sites include the Hagia Sophia church, now a museum, the Galata Tower, the Golden Horn, and the Sultan's Topkapi Palace.

The population of Istanbul was about 250,000 at the time of an earthquake in 1509 and about 700,000 in 1927. It was slightly more than one million in 1945, at the end of World War II, after which the city began an explosive growth. By 2012, Istanbul had grown to nearly 15 million inhabitants. The city expanded to the Asian side of the Bosphorus, where Üsküdar, formerly Chalcedon, is now a suburb. One controversial proposal of the Turkish government is to build a canal that would serve as a second Bosphorus, which could raise the population to 25 million. If this happens, Istanbul would become one of the largest cities on Earth.

Its population explosion is one of the reasons I classified this magnificent city as an earthquake time bomb. The second reason is the North Anatolian fault, which is in the Sea of Marmara, close to and south of the city.

NORTH ANATOLIAN FAULT

Istanbul is on the boundary between two continents, Europe and Asia, but the North Anatolian fault is a different kind of boundary between the Eurasian and Anatolian tectonic plates (Figure 15.1). The North Anatolian fault has had many earthquakes in the past and is now under high alert for an earthquake near Istanbul as large as the Izmit earthquake of magnitude 7.4 that struck the North Anatolian fault east of the city in 1999 and killed thousands of people, including residents of the easternmost suburbs of Istanbul. Because of the threat to Istanbul, the 1999 earthquake is now called the Marmara earthquake by the residents of Istanbul, named for the Sea of Marmara, where the epicenter of the next earthquake is expected to be located (Figure 15.1).

The Middle East is being jammed together tectonically by the northward motion of the Arabian plate against Eurasia, a motion that is expressed by the uplift of the high Caucasus mountains. However, the response to this northward motion is different in Asiatic Turkey,

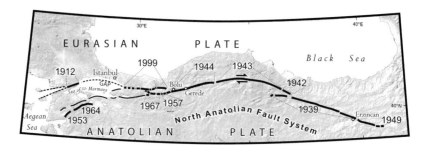

FIGURE 15.1 The North Anatolian fault in Turkey, marking the boundary between the Anatolian plate on the south and the Eurasian plate on the north. The numbers are the dates of earthquakes on the fault, emphasized by heavy lines, beginning with the 1939 Erzincan earthquake of magnitude 8 and continuing with a westward-migrating zone of earthquakes, with the most recent being two earthquakes in 1999 east of Istanbul. The 1912 earthquake farther west in the Dardanelles Peninsula is separated from the 1999 earthquakes by a seismic gap opposite the megacity of Istanbul that has not experienced a major earthquake since 1766. Source: Yeats (2012, Fig. 7.10) and Kondo *et al.* (2005).

which consists of the Anatolian peninsula between the Black Sea and the Mediterranean Sea. Instead of being forced upward as high mountains like the Caucasus, the Anatolian plate, bounded by the North Anatolian fault on the north (Figure 15.1) and the East Anatolian fault and the Cyprus tectonic zone on the south (Cyprian Arc, see Figure 14.1), is being squeezed westward toward the Aegean Sea like a giant prune seed. The north and south boundary faults have been the sources of large earthquakes in the last few centuries.

The North Anatolian fault extends west from the Anatolian peninsula into the Sea of Marmara, just south of Istanbul (Figure 15.1). Because it is underwater there, it is not directly visible, but it has been identified by offshore geophysical surveys, and it is located as accurately beneath the sea as it is on dry land east and west of the Sea of Marmara. The build-up of strain east of the Sea of Marmara translates to a slip rate on the fault of nearly an inch (22–25 mm) per year. At the Sea of Marmara, the fault divides into two zones of faulting (Figure 15.1). The northern zone passes close to Istanbul and has by far the higher slip rate, about 0.75 inch (18 mm) per year, which is a measure of its higher earthquake potential.

The long history of record-keeping includes evidence of large earthquakes throughout the history of Istanbul/Constantinople/ Byzantium, documented by the research of an engineering seismologist, the late Professor Nicholas Ambraseys of the Imperial College of London, summarized by Ambraseys (2009). An earthquake on December 14, AD 557 caused the collapse of the main dome of Hagia Sophia church. Another earthquake on January 9, 869 caused a partial collapse of the dome. On September 10, 1509, after the conquest of the city by the Ottomans, an earthquake of magnitude estimated as 7.2, described in historical documents as the Lesser Judgment Day Earthquake, struck in the Sea of Marmara, taking 10,000 lives and destroying 109 mosques and more than 1000 homes. The earthquake may have gotten its name because the Ottomans had plastered over Christian mosaics on the walls of Hagia Sophia church and converted the church into a mosque. The earthquake caused the plaster to fall away, exposing the mosaics of Christ and the saints. In addition, a minaret collapsed that had been added to Hagia Sophia after the Ottoman conquest. These "miraculous" events became widely known throughout Christian Europe, still smarting from the conquest of Constantinople by Muslims a half century earlier.

As summarized by my late friend and colleague, Professor Aykut Barka of Istanbul Technical University, and by Kondo *et al.* (2005), the most unusual sequence of earthquakes on Earth began on December 26, 1939, near Erzincan, a city in eastern Anatolia, a few months after the beginning of World War II (Figure 15.1). The earthquake was of magnitude 8, the largest ever recorded on the North Anatolian fault by seismographs. Between 20,000 and 40,000 people were killed, and more than 30,000 houses were destroyed in the early stages of a bitterly cold winter. Because of World War II, Western scientists were unable to investigate the earthquake directly, but it was studied in detail by a team led by a Turkish seismologist, Professor Ihsan Ketin of Istanbul Technical University.

Surface breaks on the North Anatolian fault during the 1939 earthquake were found by Ketin's team over a distance greater than 200 miles (340 km), with displacements as much as 25 feet (7.5 m). Three years later, on December 20, 1942, another earthquake of magnitude 7.1 ruptured another segment of the fault west of the Erzincan rupture over a distance of 30 miles (50 km), and less than a year after that, on November 16, 1943, an earthquake of magnitude 7.3 struck still farther west with a surface rupture of 175 miles (280 km). Two months later, on February 1, 1944, an earthquake of magnitude 7.3 extended the rupture another 110 miles (180 km) westward. Because World War II was still being waged, the field study of these earthquakes was limited to the work supervised by Ketin. The western progression of earthquakes could be compared to falling dominos, with each domino west of the one before it, and the earthquakes striking closer and closer to the city of Istanbul over a period of less than five years.

After this sequence of earthquakes, the war ended, and the North Anatolian fault, with its series of earthquakes studied by Ketin's team, could be visited by scientists from around the world, although there were no more earthquakes for the next 13 years. Ketin, however, published a paper in 1948 pointing out that the earthquakes his team had studied were the result of strike-slip on the North Anatolian fault, with the south side moving sideways (westward) with respect to the north side. This was the first paper pointing out strike-slip on a major earthquake fault on a plate boundary, predating by five years a similar conclusion for the San Andreas fault in California.

The westward progression of earthquakes resumed on May 20, 1957 with an earthquake of magnitude 7. Ten years later, on July 22, 1967, an earthquake of magnitude 7.1 struck the Mudurnu Valley in western Anatolia, leading to a field study by Ambraseys and his colleagues. In 1977, ten years later, two of my Turkish graduate students and I were doing detailed field studies of the Mudurnu Valley earthquake rupture. This caused great consternation among the villagers of Mudurnu Valley, who were sure that our survey, ten years after the earthquake and 20 years after the 1957 earthquake, meant that another earthquake was imminent. However, 32 years would follow the 1967 earthquake before the westward earthquake progression would resume. Because the 1957 and 1967 earthquakes were smaller, many scientists concluded that the westward-progressing sequence of earthquakes was winding down.

The Izmit, or Koçaeli, earthquake of magnitude 7.4 on August 17, 1999, now called the Marmara earthquake, disproved these ideas, which perhaps were no more than wishful thinking. The westward progression did continue, with another 100 miles (145 km) of surface rupture on the North Anatolian fault, as far west as the eastern suburbs of Istanbul. However, the surface rupture took place on a branch of the North Anatolian fault north of the Mudurnu Valley break (Figure 15.1), one that pointed like an arrow directly at Istanbul itself. In addition, the Izmit earthquake was followed less than three months later, on November 12, 1999, by the Düzce earthquake of magnitude 7.1. That earthquake, however, did not continue the westward progression but ruptured in the opposite direction, east of the Izmit earthquake and north of the earlier earthquakes of 1957 and 1967.

The death toll from the Izmit earthquake was close to 18,000, many in the easternmost suburbs of Istanbul (Figures 15.2, 15.3, 15.4). The problem was not the lack of scientific and engineering expertise, because Turkish seismologists and geologists are world leaders in earthquake science, as illustrated by the groundbreaking studies by Ihsan Ketin and by seismologists such as M. Nafi Toksöz, a Turkish expatriate who spent most of his career at MIT. The problem also was not a lack of modern building codes, again because Turkish engineers had designed their codes to deal with the earthquakes they knew would some day strike Istanbul and other urban centers. Turkish building codes are comparable to those in the United States, but their enforcement was another matter.

Despite building codes, many apartment buildings had been erected with shoddy construction to deal with the migration of people from the rural areas of Anatolia to job opportunities in the growing city of Istanbul. Overcrowding led to a demand for low-cost housing

FIGURE 15.2 A mosque and its minaret in the town of Golcuk, 60 miles (100 km) east of Istanbul, did not collapse during the 1999 Izmit earthquake, although other buildings nearby collapsed into rubble with great loss of life. Clearly, Turkish builders were able to construct safe religious buildings although not ordinary apartment buildings nearby. Source: photo by Enric Marti, from *New York Times*, permission granted from Associated Press.

at a time when inflation was high. One factor in the large death toll, pointed out by Professor Ambraseys, was a problem throughout the developing world: corruption in the building construction industry, including poorly paid building inspectors who were bribed to look the other way rather than enforce Turkish building codes (Ambraseys and Bilham, 2011). Regulations were not accompanied by effective enforcement, and there was a lack of accountability. This was illustrated by comparing the construction of apartment buildings and mosques. Figure 15.2 illustrates an example of mosques and minarets that survived the earthquake, whereas apartment buildings collapsed.

The large death toll from the Izmit earthquake led to heavy criticism of the Turkish government's lack of preparation. The

FIGURE 15.3 These buildings in Golcuk, east of Istanbul, were heavily damaged in the 1999 earthquake, but their frameworks maintained their integrity, thereby permitting many occupants to escape. Source: photo by Mustafa Erdik, Bogaziçi University, Istanbul.

difference in response of buildings is illustrated in Figure 15.4. The buildings on the lower left and in the center of the photo collapsed completely, killing everyone inside, whereas the buildings on the upper right and upper left, although damaged, allowed their residents to escape with their lives. The severely tilted buildings in Figure 15.3 were damaged, but they maintained their integrity so that most of the inhabitants survived. But the corruption problems in Turkey at the time of the earthquake, as monitored by an independent non-profit agency, Transparency International (www.transparency.org/pol icy_research/surveys_indices/cpi/2012/results) as the Corruption Perception Index (CPI), were similar to those in neighboring countries in the Middle East.

Ambraseys and his colleagues looked back at earlier earth-quakes in and near the Sea of Marmara to find a pattern of faulting and earthquakes that might forecast the next big earthquake on the

FIGURE 15.4 Buildings in Golcuk, 60 miles (100 km) east of Istanbul, largely collapsed during the 1999 Izmit earthquake. They were not reinforced against lateral shear, which caused the loss of many lives. In contrast, buildings in the right-hand background had relatively little damage, another example of differing construction standards. Source: photo by Mustafa Erdik, Bogaziçi University, Istanbul.

North Anatolian fault. That took them back to earthquakes hundreds of years ago.

On August 17, 1668, a huge earthquake of magnitude 7.9 in central Anatolia, almost as large as the 1939 Erzincan earthquake of magnitude 8, broke along the North Anatolian fault for a distance close to 250 miles (400 km). More than a half century later, on May 25, 1719, the next segment to the west ruptured in an earthquake of magnitude 7.4. This was followed by an earthquake on the next segment to the west on May 22, 1766 of magnitude 7.1 and another one on August 5, 1766 of magnitude 7.4. These earthquakes were accompanied by rupture of the entire North Anatolian fault all the way across the Sea of Marmara, whereas the 1999 earthquake left an unruptured gap directly south of Istanbul, a gap previously filled by the May 1766 earthquake. West of that gap, an earlier earthquake had ruptured across the Gallipoli (Gelibolu) Peninsula on August 5, 1912, the same location as the earthquake of August, 1766.

Using the predicted strike-slip rate of slightly less than an inch (18 mm) per year, how long would it take for the fault to accumulate enough strain to generate an earthquake greater than magnitude 7? It would take at least 200 years of strain build-up to produce an earthquake as large as the smallest earthquakes of the 1939–1999 sequence. But my Turkish colleague Celal Şengör of Istanbul Technical University cautions that one must also take into account deformation away from the main fault, which he includes in what he calls the North Anatolian Shear Zone. In addition, earlier earthquakes did not break exactly along the same part of the fault as modern ones did. Using this method would give no more than a very rough estimate of what the return time for earthquakes would be on the same part of the North Anatolian fault.

The section of the North Anatolian fault that broke in the 1999 Izmit earthquake had last broken in 1719, a recurrence interval of 280 years. That part of the fault that ruptured in the Gallipoli Peninsula earthquake of 1912 had ruptured earlier in 1766, an interval of 146 years. The fault opposite Istanbul has not generated an earthquake since May 22, 1766, a period of 246 years, shorter than the recurrence interval of the 1999 earthquake but elevated in its hazard because both sides of the seismic gap had ruptured in the twentieth century. The May 22, 1766, earthquake in the Sea of Marmara caused the collapse of houses and public buildings in Istanbul and was accompanied by a tsunami. More than 850 people were killed, a figure later raised to 4000–5000. A large mosque dedicated to Sultan Mehmet II, the Conqueror of Constantinople, was destroyed, and 100 students in a school within the mosque were killed. Most mosques and churches were damaged, as was the Topkapi Palace, requiring the sultan to live in temporary quarters until repairs were made. Contemporary European reports compared the earthquake to the great Lisbon earthquake of 1755, 11 years earlier.

Ambraseys went back further in time to a set of earthquakes in the sixteenth century, shortly after the capture of Constantinople by the Ottomans. The fault south of Constantinople/Istanbul produced an earthquake of magnitude 7.2 on September 10, 1509, the so-called Lesser Judgment Day earthquake, 257 years before the next earthquake at the same place in 1766. The 1509 earthquake was followed by an earthquake of magnitude 7.1 on May 10, 1556 on a different part of the fault. A modern earthquake has not happened there yet, and adding 257 years to 1766 brings the next earthquake to Istanbul in 2023, a decade from now. However, it could strike tomorrow or long after 2023.

THE ISTANBUL EARTHQUAKE FORECAST

We know that the North Anatolian fault in the Sea of Marmara south of Istanbul breaks in large earthquakes of magnitude 7 or larger, and we also know that the return time of these earthquakes on the same section of the fault is a few hundred years. After the 1999 Izmit (or Marmara) earthquake, separate papers were published in 2000–2004 by Tom Parsons and Ross Stein of the USGS, Aykut Barka and Kuvvet Atakan of Turkey, Aurélia Hubert-Ferrari of France, and several additional international experts. These papers contain long-range forecasts of the next earthquake near Istanbul, the most detailed and specific forecasts for any fault on Earth. (Tragically, my late friend and colleague, Professor Aykut Barka of Istanbul Technical University, involved in three of the forecasts, lost his life in an automobile accident after the forecasts were made public.) These forecasts state that there are two chances out of three that the Istanbul segment of the North Anatolian fault will be the source of a large earthquake in the next 30 years! Another forecast gave the 30-year probability of an earthquake of magnitude 7 or larger as 41%. These are scary numbers!

These are *probability forecasts*, comparable to a climate forecast of the probability of another hurricane the size of Hurricane Katrina striking New Orleans in the next 30 years. Probability forecasts affect insurance rates and should influence the enforcement of building codes, although they would not affect your vacation plans. These probability forecasts have affected decisions by the Turkish authorities in planning for the next earthquake in Istanbul, as discussed below. Two chances out of three of a large earthquake in the next 30 years have major implications for the safety of the growing population of Istanbul and for future development of the city.

Is the probability forecast an earthquake prediction? No, it is not. Scientists are trying to figure out how to predict earthquakes, or at least some types of earthquakes, but no luck yet (see Chapter 4). Some say that the structure of the Earth is too complicated for any reliable prediction, and one must consider the social upheaval if a future prediction is wrong. However, the question we scientists are most frequently asked (as we should be asked) is: "When's the next big earthquake?" Our answer, sadly, is still "We don't know."

ISTANBUL'S RESPONSE TO THE FORECASTS

However, the probability forecasts have been enough for the authorities in Istanbul to take action, building on their cadre of world-class experts in earthquake engineering, seismology, and urban planning. (This section has benefited from the work of Mustafa Erdik of Kandilli Seismic Observatory and Bogaziçi University, and Mehmet Celebi and Ross Stein of the USGS; see Erdik's paper in *Science*, 2013.) An earthquake in eastern Turkey on October 23, 2011 of magnitude 7.1 that killed 604 people and left 60,000 homeless added to the urgency to respond to the earthquake forecasts.

Unlike the authorities in several of the cities labeled earthquake time bombs in this book, including Caracas, Tehran, Kabul, and where I live in western Oregon in the United States, the Istanbul Metropolitan Municipality (which I refer to here as Metro Istanbul) and the Governorship of Istanbul decided to take action. The high probability of an earthquake is combined with the fact that the economy of greater Istanbul is more than 40% of that of the entire country, and the population of Istanbul comprises 20% of the population of

Turkey. This means that the next Istanbul earthquake could be cata-strophic for not just Istanbul but for the entire country of Turkey.

First, Metro Istanbul charged the four major universities of the Istanbul region to recommend a master plan of action, which they presented to the government in 2003. The study by the universities first assessed the cost of the 1999 Marmara earthquake east of Istan-bul: 113,000 buildings destroyed, 264,000 damaged, up to 600,000 homeless, and a cost of $10–15 billion, 5–7% of Turkey's gross national product. Communications networks failed, and first aid and rescue operations were disorganized, hampered by the inefficiency of governmental bureaucracy. The cost of a future earthquake (Istanbul earthquake, producing a direct hit on the city) of magnitude 7.5 was estimated as $30–40 billion in losses and up to 30,000 killed. The governments, including Metro Istanbul and the national government, with the support of the population, agreed that these losses under present conditions would be unacceptable.

The Governorship of Istanbul, through its Disaster Manage-ment Authority, used its Earthquake Master Plan to apply for and receive a loan of €860 million in credit from the World Bank, European Investment Bank, and European Union Development Bank. After the Marmara earthquake, the government also started collecting tempor-ary taxes to help offset the costs of the next disaster. Pilot projects were started by Metro Istanbul in suburbs where the danger was considered especially high, including a study of the suburb of Zeytinburnu. The Zeytinburnu project identified at least 2295 out of 16,031 buildings in that suburb as being at extremely high risk of heavy damage by an earthquake of magnitude greater than 7. During the pilot project, two buildings in Zeytinburnu actually collapsed *without an earthquake*, evidence for the need to upgrade the quality of construction throughout the city.

The overall project under the direction of the Governorship of Istanbul was called the Istanbul Seismic Risk Mitigation and Emer-gency Preparedness Project (ISMEP), to be carried out in the period 2006–2014. An important component of the project was and is public awareness, including organizations down to the level of neighbor-hoods and families to swing into action after the expected earthquake. There were 500,000 brochures prepared for residents of Istanbul to bring people up to date on the project and involve them in steps they can take to protect their families and local businesses.

Another component is the enforcement of building codes to overcome the corruption that has doomed similar projects in the past. The Turkish building code has been updated, and 30,000 civil engin-eers are being trained to inspect buildings, including upgrades of historic buildings. This requires a major cultural change in the con-struction industry to engage builders to build safely. Buildings must be retrofitted against earthquakes, and this is costly. Public buildings (hospitals, schools, administrative buildings) are being retrofitted through a loan of €1.3 billion. About 12% of 308 hospitals and 35% of 1783 schools have already been retrofitted under the ISMEP program.

Even though the project is still underway, it has caused Trans-parency International in 2012 to upgrade the Corruption Perception Index for Turkey to 49, higher than the ranking of most neighboring countries, including Greece, which has a rating of 36. In comparison, Iran ranks 28, Syria is 26, and Afghanistan is 8, 174th out of 176, the total number of countries evaluated. The project is still a work in progress, in that the future of the country and the safety of its citizens depend on the honesty of its construction industry and of its building inspectors and the determination of decision makers in Istanbul.

The project has not been without problems. It has been criti-cized for its "top-down" administration. The increase in taxes has been unpopular. Another problem is with building owners: Who pays for retrofitting? For example, a seismic retrofit of a building does not necessarily increase its value, and so building owners resist having to spend money without an immediate pay out. One plan is to reduce the number of "pancaked" buildings, where total collapse results in the deaths of all the people inside. It would be preferable to give top priority to retrofits that result in greatly reduced loss of life, even if the building is damaged beyond repair. Examples are illustrated in the tilted buildings in Figure 15.3. Firms supervising construction are required to have liability insurance, and it is difficult to estimate what premiums to charge to cover the insurance or reinsurance company in the event of a destructive earthquake. As a result, the insurance requirement for private housing has not been fully implemented.

The World Bank project has led to a government-sponsored Turkish catastrophic insurance pool (TCIP), to transfer the financial burden of replacing earthquake-damaged housing from the government to international reinsurance. The insurance pool is patterned after the California Earthquake Authority (CEA) and the New Zealand Earthquake Commission (EQC). All existing structures are required to join this pool, although this has met with resistance from building owners and landlords. At present, the compulsory earthquake insurance pool is probably inadequate to cover the costs of the expected earthquake. It will be necessary to use public funds for many retrofits because the owners will simply not have the resources to cover the costs of earthquake recovery.

As reported by Erdik in 2013, the Turkish government is implementing a 20-year urban renewal program costing $400 billion, which will include the demolition of seven million residential units, which will greatly reduce the loss of life if the next earthquake does not strike until after the urban renewal program is completed.

The terminal building of Istanbul International Airport, damaged during the 1999 Marmara earthquake, has been the subject of a separate earthquake-related project. A two million square foot building was redesigned using *base isolation*, the world's largest structure using this principle. The base isolation principle works like shock absorbers in a car, so that the structure will not collapse in the next earthquake.

In my view, the high caliber of Turkish seismologists and engineers has made this project possible. This has led to the involvement of GeoHazards International, an American nonprofit firm focusing on earthquake hazards, in working with Turkish seismologists and engineers to upgrade the building stock in Istanbul and elsewhere in the country, with support from USAID and the Earthquake Research Institute. It is also clear that the decision makers at Metro Istanbul care about the safety of their citizens. Many other earthquake time bombs could and should benefit by learning from the Istanbul project, which will be tested in the near future by an earthquake of magnitude greater than 7. I wish them well, and I hope that the next earthquake gives them their full 20 years to complete their retrofit. I hope also that the corruption problem that contributed so much to the losses in the 1999 earthquakes has been overcome. Time will tell.

I dedicate this chapter to a Turkish geologist who devoted his life to understanding the hazards from a rupture of the North Anatolian fault, and who participated in the probability forecasts that led to the present project to strengthen Istanbul against the next earthquake: the late Professor Aykut Barka of Istanbul Technical University.

16 Tehran: the next earthquake in the Islamic Republic of Iran?

Ever since the 1979 Islamic revolution and the return of Ayatollah Khomeini, Tehran, the capital city of Iran, has been much in the news because of the anti-Western theocratic rule of its current Supreme Leader, Ayatollah Ali Khamene'i, and the behavior of the previous President of Iran, Mahmoud Ahmadinezhad, principally because of worldwide concern that the Islamic Republic of Iran is developing nuclear weapons to use against Israel. The comings and goings of several Iranian presidents over the past few decades, whether hard-liners or moderates, have been against the backdrop of the Supreme Leader, supported by the Revolutionary Guards, who is the final authority. Today's president, Hassan Rouhani, appears to be in favor of *rapprochement* with the West, and a diplomatic solution is now being considered by Iran and Western nations.

However, the focus here is on the potential for a devastating earthquake that would destroy much of Tehran and have a major effect on the politics and balance of power of the Middle East. Tehran's earthquake potential has largely gone unrecognized by the media, the general public, and, with one notable exception, Iran's leadership.

Iran is an Islamic country but with some qualifications. First, the country is home to significant minority populations of Zoroastrians, Jews, and Christians. Second, the Iranians, historically called Persians, are not Arabs; most of them speak their own language, Farsi.

The Persian people have a much older cultural history than the Arabs who swept through the country in the seventh century AD. The first Persian empire was that of the Medes in the seventh century BC, followed by the Achaemenid Empire of Cyrus the Great and Darius. That empire in the sixth century BC was the first world empire, extending from the Indus River of modern Pakistan to the Nile River of Egypt and the Aegean Sea, where the Persian wars against Greece included the burning of the city of Athens in 480 BC. The Achaemenids were conquered by Alexander the Great, who established the Greek-influenced Seleucid Empire, which was itself overthrown by another Persian dynasty, the Parthian Empire, in the third century BC. The Parthians and, later, the Sassanians were in frequent conflict with Rome, but Rome never conquered Persia.

However, the wars with Rome weakened Persia, and as a result the Sassanian Empire was overrun by the Arabs, revitalized under the Prophet Muhammad, in AD 636–642. Over a period of about two centuries, the Persian Zoroastrian religion was supplanted by Islam. A large part of what is historically known as the Islamic civilization is based on its Persian cultural and political heritage.

Although Tehran is a megacity, that has not always been the case. Indeed, Tehran is the 32nd capital city of Iran, established in 1795 by Qajar kings. Tehran was not a major center of culture in classical times, although prior to the Arab conquest, the modern city of Ray, now a suburb in southern Tehran, was an important center called Rhagae. Earlier capitals, such as Tabriz in northwest Iran, have a long history, and Tabriz has itself had a history of destructive earthquakes, as discussed in a later section.

It is a mistake to refer to Iran as a developing country. During medieval times, Persia was a center of science and culture, surpassing Europe at that time and comparable to the Middle Kingdom of China. Its culture and traditions predate those in the West. In the modern era, Iran has produced scientists of the first rank, and its universities are well regarded. Educated Iranians who have chafed under religious rule of the ayatollahs since 1979 have fled the country, producing a brain drain in Iran, but benefiting the democratic societies of the West.

I have worked with an eminent Iranian scientist, Manuel Berberian (Berberian and Yeats, in press), who, through the Geological Survey of Iran, upgraded Iran's understanding of the earthquake faults that are found throughout the country. Berberian, who holds a PhD from the University of Cambridge and comes from an Armenian Christian family, has contributed to the seismic and tectonic studies of Iran for more than four decades. However, he has chosen to live in the United States since the 1990s because of unsuitable conditions for his family in Iran. His departure was a major loss to earthquake research in Iran, although he has continued to study Iranian earthquakes since leaving the country. His lifelong contribution to earthquake science was featured in a symposium in his honor at the 2013 annual meeting of the Geological Society of America in Denver. Participating in the symposium was a large number of scientists not only from the United States and Europe, but also from Iran, an important milestone in overcoming political obstacles to attend international conferences in a truly scientific spirit.

Tehran lies at the southern foot of the Alborz (Elburz) Mountains (Figure 16.1), which are crowned by a dormant volcano, Mt. Damavand, 18,406 feet (5610 m) high, the highest mountain not only in Iran but in the entire Middle East. The Alborz Mountains frame the southern end of the Caspian Sea and are subject to earthquakes. The most recent large earthquake in the Alborz Mountains was the Rudbar earthquake of magnitude 7.3 on June 20, 1990, southwest of the Caspian Sea and northwest of Tehran. More than 40,000 people lost their lives in that earthquake. Three cities and 700 villages were destroyed. The faults that were the source of the earthquake were not previously known to be active, and the area most strongly affected by the earthquake was in a seismic gap, in which a major earthquake had not struck in many centuries.

The most recent official population estimates for Tehran are more than 8.4 million, four times its population in 1960. Unofficial population estimates are higher, up to 12 million people. The sources of this dramatic increase in population are Iranian refugees from the Iran–Iraq war of 1980–1988 and Afghan refugees from the wars in Afghanistan, including the Soviet occupation and the civil war with the Taliban that followed. Another, and probably more important, cause of the population increase is the movement from the countryside, where there are few jobs, to Iran's largest commercial city.

FIGURE 16.1 View north from north Tehran to the snow-capped Alborz Mountains, which contain several active faults, posing a hazard to Iran's capital city. The North Tehran thrust at the foot of the Alborz is active. Source: photo by R. Q. M. Taban of Tehran, courtesy of Manuel Berberian, from a paper in preparation for the Geological Society of America by Berberian and Yeats.

Why is Tehran located where it is, at the foot of the Alborz Mountains (Figure 16.1)? The mountains block the southward flow of moist air from the Caspian Sea, so that the Tehran region, like most of the Iranian Plateau, is a desert. Tehran, like many other cities of central Asia, developed where it did because its inhabitants could take advantage of rivers flowing from the mountains into the parched plains to the south. Since ancient times, farmers dealt with their desert climate by constructing ingenious underground tunnels called *qanats* that channel the water and permit its use for irrigation. Tehran has long since outgrown its qanat system, but qanats are still used in parts of the desert regions of Iran.

The problem for Tehran is that the southern edge of the Alborz Mountains, as well as more interior parts of the range, is formed by active faults that over many thousands of years have uplifted the mountains, with the uplift accompanied by earthquakes. The mountains provide a bounty of fresh water, but at a terrible cost. A rapidly growing city with poor construction practices located at the foot of a mountain range because of its access to water must face severe damage and destruction when one of the faults forming the mountain front releases its built-up strain as a large earthquake.

Iranian scientists, including Manuel Berberian, have studied the active faults in the Alborz Mountains and identified several that are close to Tehran, principally the North Tehran thrust fault and several additional faults (Figure 16.2). The Mosha fault was the source of an earthquake of magnitude 7.1 on March 27, 1830, the most recent damaging earthquake to strike the Tehran metropolitan region. Other active faults, including the Davudieh and Mahmudieh faults, lie south of the Alborz Mountain front within the city itself. These faults within metropolitan Tehran are now covered by buildings where thousands of people live and work.

The faults beneath and adjacent to Tehran have not shown evidence for a major earthquake in several centuries (Ambraseys and Melville, 1982). This includes the large North Tehran thrust fault, which forms the Alborz range front adjacent to the city (Figure 16.2). Paleoseismic backhoe trenching by a French–Iranian team gave evidence for a medieval earthquake on the North Tehran fault of magnitude 7.2 in AD 1177 and an earlier earthquake of magnitude 7.6 during the interval 312–280 BC, while the region was part of the Seleucid Empire. The ancient city of Ray, now a southern suburb of Tehran, has a much longer historical record than Tehran itself. Ray experienced at least two earthquakes, including earthquakes in 312–280 BC and AD 1177, but none since. The 1177 earthquake at Ray may have been the same one documented on the North Tehran thrust by paleoseismic trenching.

Berberian interprets the absence of earthquakes in recent times as evidence that Tehran is in a seismic gap. That is to say, strain continues to build-up on faults in and near Tehran, soon to be released

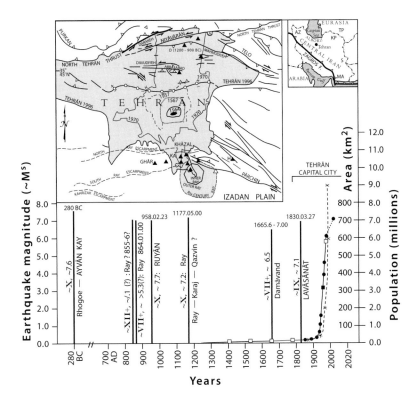

FIGURE 16.2 Map of Tehran showing increase in metropolitan area from 1564 to 1996, the most recent boundary showing modern Tehran in gray. Solid lines with arrows mark strike-slip faults, lines with triangles mark thrust faults. Dashed broken lines (escarpments) may mark faults not reaching the surface, as does the line marked by open circles. Solid triangles are archaeological sites. The diagram below shows historical earthquakes with magnitudes where known; modified Mercalli intensities with Roman numerals for pre-instrumental earthquakes. The most recent damaging earthquake in the Tehran metropolitan area struck in 1830, indicating that Tehran is currently in a seismic gap. The population curve shows the major growth of Tehran in the last century; it may underestimate the actual growth. Source: from a paper in preparation by Manuel Berberian and R. Yeats for publication by the Geological Society of America.

in a destructive earthquake, repeating the disaster of the 1990 Rudbar earthquake, but in a much larger city. Will Iran's leaders, including the current government of moderate President Hassan Rouhani, rise to the challenge and strengthen their capital against earthquakes *before* the next earthquake strikes? Or will they take the fatalistic approach of the Muslim theocratic leadership about the possibility of a destructive earthquake: If it happens, *insh'allah*, it's the will of God. The answer to this question is not yet at hand, although the previous president, Mahmoud Ahmadinezhad, proposed moving millions of people away from Tehran's danger zones, as described below.

On December 22, AD 856, an earthquake ruptured a fault in the Alborz Mountains east of Tehran, near the cities of Damghan and Qumis. Its magnitude was formerly estimated as large as 7.9 or 8, the greatest historical earthquake to strike the Alborz region and the sixth most disastrous earthquake in human history worldwide. An estimated 200,000 people were thought to have lost their lives. But a reanalysis of historical documents by Berberian has shown that the supposed size of this earthquake was due to its being the source of several earthquakes at about the same time. The 856 earthquake had a magnitude greater than 7, resulting in thousands of deaths – a major disaster to be sure, but not as high as magnitude 7.9, and losses in the tens of thousands, but not in the hundreds of thousands.

Berberian's historical analysis is important to Tehran because if the 856 earthquake had a magnitude of 7.9–8, this would be used as the worst-case deterministic scenario earthquake for Tehran, meaning that any structures in Iran, especially the country's nuclear power plants, must be engineered against an earthquake of this magnitude. An earthquake that struck the North Tehran thrust fault and the Niavaran fault in 312–280 BC that might have been as large as magnitude 7.6 is another candidate for worst-case scenario, but historical documents for this earthquake do not favor an earthquake this large. However, even if the worst-case Tehran earthquake were to have a magnitude as low as 7.2, the large population of the Tehran metropolitan area (Figure 16.2) would lead to a catastrophic loss of life, probably larger than any earthquake in Iran's history.

Even though Tehran is in many ways a modern city and has modern building codes, the standards of its construction industry are questionable and worrisome, meaning that shoddy building practices are tolerated and overlooked by corrupt building inspectors. The global organization Transparency International (www.transparency. org/policy_research/surveys_indices/cpi/2012/indetail4) rates all countries on Earth using a Corruption Perception Index (CPI); in 2012, Iran rated a low score of 28 out of a possible 100. Building inspectors are poorly paid and accept bribes, so that building codes are not enforced. Meanwhile, the demand for housing in Tehran's expanding population is too great. The deaths of hundreds of thousands of people would lead to severe social turmoil for any Iranian government. Tehran, because of its large population living among active earthquake faults, is therefore a political as well as an earthquake time bomb.

But wait a minute! The former President of Iran, Mahmoud Ahmadinezhad, offered a solution. In a press conference, he acknowledged the disastrous earthquake of 1830 on the Mosha fault and the presence of active faults within the city of Tehran. The president's solution: move several million Tehranis out of the city to locations elsewhere in Iran, reversing the trend of increasing Tehran's population in recent decades. It was obvious to me that the president had seen a fault map of Tehran, very likely prepared by Berberian in 1983, when he was with the Geological Survey of Iran (Berberian and I published an updated version of this map in 1999). Ahmadinezhad's proposal might seem bizarre, except for the fact that the capital of Iran has moved many times in the past, and if there are places safer from earthquakes than Tehran, perhaps the country should take advantage of the better knowledge of Iranian earthquake hazards brought to light by Berberian and his Iranian colleagues.

The motivation for Ahmadinezhad's proposal is controversial. To me, he was the first political leader on Earth to make a serious proposal to alleviate the risk to inhabitants of a major earthquake-prone city by moving them, along with government offices, to safer areas. His proposal was supported by a prominent geophysicist, Professor Bahram Akasheh, of Tehran University. But to many other Iranians, the motivation was rhetorical (without a practical way to implement the suggestion), populist (appealing to public opinion), or political (the people that the president would have moved away from earthquake danger live in northern parts of Tehran, where he had faced political opposition, including riots, before his controversial re-election in 2009).

What areas of Iran would be relatively less earthquake-prone and suitable for such a massive re-settlement of people? As pointed out above, Iran has had other capitals, and these have also been the sites of earthquakes. One former capital, Tabriz, experienced two major earthquakes on the North Tabriz fault north of the city, including an earthquake of January 8, 1780, with its magnitude greater than 7.4, and an earlier one on a different part of the fault on April 26, 1721 of magnitude at least 7.3, which caused 77,000 deaths. A still-earlier earthquake struck Tabriz on November 4, 1042, of magnitude 7.3. Recently, a pair of moderate-sized earthquakes of magnitude 6.3 and 6.4 struck a mountainous region northeast of Tabriz on August 11, 2012, taking the lives of more than 300 people.

A contemporary Iranian scholar, Yah'ya Zaka, has given a detailed historical discussion of earthquakes in Tabriz in a Persian book titled *Zamin larze-ha-ye Tabriz* (*Earthquakes of Tabriz*, 1989). The historical record dates back to the ninth century and includes nearly 120 notable earthquakes up until the end of the twentieth century.

My suggestion to the Iranian government is to reinforce buildings in Tehran against earthquakes. This is based on the fact that Iran, a major oil producer and member of the Organization of Petroleum Exporting Countries (OPEC), is not a poor country. However, economic sanctions and political isolation have been imposed on Iran by the international community because it has not been forthcoming in allowing international inspection of its nuclear facilities. The sanctions have curtailed Iran's ability to export its oil, causing severe inflation and a shortage of hard currency. The Iranian regime's anti-American and anti-Israeli rhetoric and its lack of openness, leading to the international fear that Iran is developing nuclear weapons to be

used against Israel, have damaged what otherwise would be a prosperous economy. Accordingly, the strengthening of Tehran against its
inevitable next major earthquake is left undone.

In June 2013, Iranian voters overwhelmingly elected, from
among a handful of approved candidates, a relatively moderate cleric
as president: Hassan Rouhani, who campaigned on the Iranian economy and advocated a less-confrontational approach to the international community. It remains to be seen if he follows up on the
proposal of moving the capital or advances any other solution to
mitigate the tragedy of a catastrophic earthquake in Tehran. Preliminary discussions between Iran and Western countries in Geneva have
led to a temporary agreement for six months while both sides hammer
out a more permanent solution. Rouhani and President Barack Obama
exchanged letters after his election, and they talked briefly on the
telephone in 2014 when both of them came to New York to address
the United Nations. Although negotiations are underway, there are
hard-liners in Iran and in the US Senate, as well as the Prime Minister
of Israel, who oppose these negotiations.

Resolving the political controversy about inspection of Iran's
nuclear facilities is important because Iran's own scientists and engineers are fully capable of advising the government in dealing with
Tehran's earthquake problem. The Nuclear Research Center and the
Applied Physics Research Center in Tehran are both located close to
active faults. Iran's recent response to earthquakes in April 2013 near
the Bushehr nuclear power plant on the Persian Gulf and a magnitude
7.7 earthquake on the subduction zone at the Iran–Pakistan border
north of the poorly populated Makran coast has been informative
and professional. Civil engineers in Iran are able to design structures
that can significantly lower the loss of life from a major earthquake.
Iranian political leaders need to listen to their own highly qualified
engineers, seismologists, and geologists, inspect and bring up to international standards the existing buildings in Tehran, and require new
construction to adhere to these standards, in addition to having them
inspected by honest, qualified, well-paid building inspectors. In this

way, the Iranian government would be viewed as one that respected the lives of its own citizens.

Iran need look no further than its Muslim next-door neighbor, Turkey, itself under a probabilistic forecast of a major earthquake in Istanbul in the next few decades, as described in the preceding chapter. In contrast to Tehran, Istanbul has taken the earthquake forecast seriously and, assisted by loans from the World Bank and other international institutions, is informing its citizens of the earthquake threat and retrofitting dangerous buildings in the next 20 years. Berberian and I hope that Iran's leadership, including President Hassan Rouhani, is able to focus on this hazard and retrofit buildings in Tehran against earthquakes.

It is disappointing that in a city with as many earthquake faults as Tehran, so few paleoseismic trenches have been excavated on the North Tehran thrust, and some of these were done as projects partly funded by Western research laboratories. These trenches were done for academic research, but future paleoseismic trenching should be paid for by the Government of Iran, using funds generated from the export of oil and gas accompanying the partial lifting of sanctions. Iranian scientists are qualified to conduct paleoseismic studies, although the assistance and collaboration of foreign scientists from France, Italy, and Great Britain have been welcomed in the past. A series of paleoseismic trenches on all active faults in Tehran and Ray would allow the Iranian government to follow Istanbul's lead and estimate the probability of future earthquakes and thereby allocate priorities for upgrading Tehran's buildings (and making a cleaned-up construction industry responsible for strengthening buildings).

This would be of great value to the private insurance and reinsurance industries, which, in contrast to Istanbul, at present do not operate in Iran. A study recently published by the Insurance Research Center of Tehran showed that a majority of Iranians believe that earthquake damage should be dealt with by the government, not by the private sector. There is interest in coverage by private insurers, but for this to take place, an improvement of construction quality is

FIGURE 16.3 Earthquake damage in the 2003 Bam earthquake in southeastern Iran. Much construction in Bam, including an ancient citadel, uses mud bricks, which tend to collapse even in a moderate earthquake. Source: photo by James Jackson, University of Cambridge.

necessary. At present, estimates of building damage and losses of life are too high for the private sector to operate.

Iran and Turkey are neighboring Muslim countries under threat of earthquakes. Why has Turkey been proactive in addressing the earthquake threat to Istanbul while Iran has been less responsive for its capital of Tehran? One reason is that Turkey has been struck by several damaging earthquakes in recent decades, including two in 1999 and one in 2011, so that earthquake awareness is very high, especially in Istanbul. But the Alborz Mountains have also been struck by major earthquakes, including the 1990 Rudbar earthquake west of Tehran.

Tehran's most recent damaging earthquake was in 1830, almost two centuries ago, and this, plus mismanagement and the effect of the sanctions, may have put a major earthquake response for Tehran out of the question. However, the length of the seismic gap at Tehran should be troubling to the authorities. Another factor is that both the historical earthquake record and paleoseismic trenching for Istanbul have led to a probabilistic earthquake forecast that is as high or higher than any other city on Earth. If the Tehran metropolitan area were to undertake the paleoseismic study advocated here, Iran's leaders could get an answer as definitive as that for Istanbul within a few years.

A comparison with another part of Iran is instructive. In 2003, the desert city of Bam in southeastern Iran was destroyed by an earthquake of magnitude 6.6, taking the lives of more than 40,000 people and damaging part of the 2000-year-old Bam Citadel, a Silk Road site. One of the reasons for such a high death toll is the presence of mud-block construction that is unreinforced against seismic shaking (Figure 16.3). Buildings with this type of construction, common throughout much of Central Asia, tend to collapse in an earthquake (see Chapter 17 on Kabul). A second reason was that the Iranian government and the local population were totally unprepared. Active faulting at Bam had been recognized by Iranian geologists as early as 1976, although the source fault for the 2003 earthquake had not been recognized in advance.

In contrast, an earthquake of the same magnitude and at the same time in the California Coast Ranges took the lives of only two people!

17 Kabul: decades of war and Babur's warning

We headed north from Kabul on a warm day in July 2002, on our way to Baghlan Province and the town of Nahrin, which had been devastated by a moderate-magnitude earthquake four months earlier. In the SUV belonging to Shelter for Life International were Chris Madden and I from Earth Consultants International and Harry van Burik and Lee Molini from Shelter for Life. We were there to investigate the earthquake, working with Shelter for Life, an NGO (nongovernmental organization) that had been funded by the US Agency for International Development (USAID) to help Nahrin rebuild.

At first, the countryside looked very much like Pakistan, where I had been working on earthquake projects for nearly 25 years. But there was a difference. We passed the hulks of burnt-out tanks, and our driver pointed to little painted stones along the side of the road, marking the location of land mines. Don't go there.

The previous October, nine months earlier, Afghanistan had been attacked by its most recent invader, the United States, operating as part of NATO. In September 1996, five and a half years before, Taliban insurgents had captured Kabul and installed a rigid form of Islam, closing girls' schools, holding public executions in the stadium, and hanging the previous pro-Communist leader, President Najibullah. The Taliban had provided a base for Osama bin Laden to plan the al-Qaida attacks of September 11, 2001, against the United States. The Taliban refused to turn over bin Laden and other al-Qaida operatives to the United States, and a month later the Americans launched their offensive.

The Taliban had never completely conquered Afghanistan. The northern part of the country is dominated by minorities, principally

Tajiks and Uzbeks, who speak different dialects and follow a different brand of Islam than does the Taliban, dominated by the Pashtuns. They formed the Northern Alliance, a loosely organized group of militants who kept the Taliban at bay until 2000, when a Taliban offensive captured the city of Nahrin and drove its inhabitants into the Hindu Kush Mountains to the east. The American invasion was coordinated with the Northern Alliance and made taking over the country, including the area around Nahrin, much easier than it would have been otherwise.

Our route took us over Salang Pass in the Hindu Kush Mountains, and we crossed bridges that had been heavily damaged during the Taliban invasion and were now being rebuilt by Swedish engineers. On the north side of the mountains, we reached the city of Baghlan, capital of the province, and turned east on a rough road to the city of Nahrin, at the foot of the Hindu Kush Mountains. There we found a city that had been devastated, not by war but by the earthquake. Twelve hundred people had lost their lives out of a population of 50,000–60,000 in Nahrin and surrounding villages, and 20,000 families suffered a major impact from the earthquake, which turned out to be the most deadly earthquake anywhere on Earth in 2002.

The earthquake was of only moderate size, with a magnitude of 6.1. The large number of fatalities was caused by the type of construction: mud-blocks mortared together with mud (Figure 17.1). Although most houses had flat roofs, consisting of 10–12 inches of dried mud supported by wooden beams, the mud-block walls were not braced against horizontal shaking, and the roofs simply collapsed, killing everyone inside. In some cases, only the wooden doors and their door frames remained standing; the surrounding mud-block walls had collapsed around the door frames. A major part of Shelter for Life's goal was to show the locals how to rebuild, using local materials but including reinforcing against earthquakes, which would allow people time to leave their houses safely during an earthquake.

The earthquake struck at 7:26 p.m. local time, and most families had been preparing or eating their evening meal. Most of the

FIGURE 17.1 The town of Nahrin, Afghanistan, destroyed by a moderate-size earthquake in 2002. Mud-block walls disintegrated into rubble, leaving wooden timbers formerly covering roofs collapsed around them. The earthquake resulted in 1200 fatalities, the greatest loss of life of any earthquake in 2002. Source: photo by Chris Madden, Earth Consultants International.

outlying settlements were single-family units. When we arrived, larger families had already started repairing their houses, re-using the wood-beam roofs and wooden doors, but where the families were small or very poor, there were too few people to do the repairs, and those settlements simply remained abandoned. In some cases, the inhabitants of those settlements had all been killed, and we found ourselves documenting a tragedy.

Our interaction with the local Tajiks was unforgettable because their hospitality is representative of a large part of Central Asia. We were welcomed wherever we went. Families that had been devastated by the fighting with the Taliban had been subjected to further destruction by the earthquake, but still the elders shared with us the little they had left. They offered us tea and biscuits and would not take no for an answer. I have warm memories of our interaction with those wonderful people. They deserve better living conditions and a better existence. With NATO leaving soon, I hope they do well.

KABUL AND BABUR'S WARNING

We returned to Kabul, with its yellow taxicabs, outdoor markets, traffic policemen with white gloves, and souvenir shops on Chicken Street, but again there was a big difference from Pakistan. Many of the buildings had been damaged by the fighting, but people were living in the ruins, which looked as if they might collapse even in a small earthquake. Many people had become refugees from the previous invasion by the Soviets in 1979. Two to three million had fled to refugee camps in Pakistan and another 1.5 million had gone to Iran. They were now returning home, but there was no place for them to live, so they occupied partly destroyed buildings. If they were able to construct their own housing, they used the same mud-block construction that had failed the people of Nahrin in the 2002 earthquake (Figure 17.2).

The tragedy of the Afghan people is that they have suffered one invasion after another, starting with the Persians, who were conquered by Alexander the Great. Two thousand years and many invasions later, the Mughal emperors of India held sway, and in the nineteenth century the British attacked from India. At that time, Afghanistan and adjacent parts of India were a chessboard for the Great Game being played between Russia and Great Britain. The Russians wanted to gain a warm-water seaport, and the British fought three wars involving Afghanistan to keep that from happening. The British were successful against the Russians, but the Afghans, not having been consulted about the Great Game, had had enough of the British, and so they were kicked out of the country, back to India, in the early years of the twentieth century.

With the continuous destruction of war spanning 2500 years, it is hard to remember that Kabul also has a major earthquake problem. The Chaman fault, just west of Kabul, forms the western boundary of

FIGURE 17.2 Slums of Kabul, Afghanistan, dominated by buildings of mud-block construction which would fail during even a moderate earthquake. Kabul lacks precautions against earthquakes, even though a major plate-boundary fault is nearby to the west. Source: photo by Anthony Crone, USGS.

the Indian tectonic plate, which is colliding with Asia at a rate greater than 1.5 inches (2–4 cm) per year, uplifting the Himalaya, the world's highest mountain range. The fault gets its name from the Pakistani border town of Chaman, which was destroyed by a magnitude 6.5 earthquake on December 20, 1892, resulting in the town being relocated west of the fault (Figure 17.3).

The major earthquake that should be a wake-up call to Kabul struck nearly 500 years earlier, on July 6, 1505 (Figure 17.3), when the Mughal emperor Babur was in Kabul preparing for an attack on the city of Kandahar, far to the south. The earthquake caused great damage to the Bala Hissar fortress, as well as many other buildings. Severe disruptions of the ground were probably the result of landslides.

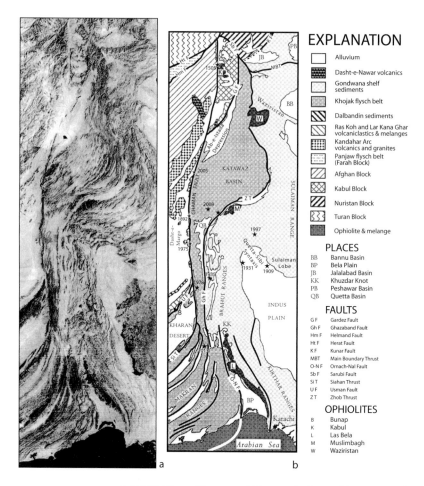

a   b

FIGURE 17.3 LANDSAT satellite map of Chaman fault in Afghanistan and western Pakistan, marking the tectonic boundary between the Indian plate on the right and the Afghan block to the left, part of the Eurasian plate. Dates close to the fault mark earthquakes, with epicenters marked by stars: 1892 marks the Chaman earthquake; 1505 marks the Kabul earthquake north of Kabul; and 2013 marks the Awaran earthquake of magnitude 7.7 on September 24, 2013. Source: modified from Lawrence *et al.* (1992) and Yeats (2012, Fig. 8.3).

At the time of the earthquake, Babur himself was ill, and during his recovery to full health, he spent a month repairing the earthquake damage to the fortress and writing his account of the earthquake. Contemporary stories suggest that there were at least 25 miles (40 km) of surface rupture, with vertical uplift of at least 10 feet (3 m). It is difficult to determine the magnitude of the earthquake, but most experts estimate its magnitude as 7.3, which is much larger than the better-documented Chaman earthquake of 1892 and many times larger than the Nahrin earthquake of 2002 that had killed 1200 people. The epicenter of the 1505 earthquake was probably a few miles north of Kabul. However, these conclusions are based on little contemporary evidence aside from the description by Babur. Study of the Chaman fault earthquake of 1505 is complicated by the occurrence of a powerful earthquake in the Himalaya that same year (see Chapter 18).

Satellite records from the Global Positioning System (GPS) suggest that the plate-boundary Chaman fault is storing strain at a rate of more than an inch (2–4 cm) per year. At this rate, in the five centuries since the 1505 earthquake, the fault has accumulated enough strain to slip more than 40 feet (12 m), which is enough to generate an earthquake at least as large as the 1505 earthquake and possibly larger, unless some of the strain has been relieved by fault creep. The southern end of the Chaman fault system in Pakistan sustained an earthquake of magnitude 7.7 in September 2013. The social consequences of a large earthquake would be catastrophic, vastly more severe than in 1505, when Kabul was a relatively small garrison town guarded by the Bala Hissar fortress, and worse than the 2013 earthquake of magnitude 7.7 in Pakistan to the south. Kabul now has a population of more than 3.5 million people, including many refugees living in mud-block houses similar to those that collapsed in the Nahrin earthquake of magnitude 6.1.

In my opinion, a repeat of the 1505 earthquake could kill more people than all the wars in Afghanistan fought over the past 30–40 years!

WHOSE PROBLEM IS IT, ANYWAY?

Is this future earthquake a concern to the Americans, NATO, or the Afghan government? Afghanistan is a poor, undeveloped country, but it is not Haiti. Since 2001, money and support have been pouring into the country. In the past 11 years, the Americans and their NATO allies have spent nearly $400 billion in Afghanistan. How much of this has gone into strengthening the Afghan economy and people to deal with the next inevitable earthquake?

USAID, which paid for our investigations at Nahrin, has funded the US Geological Survey (USGS) to prepare an active fault map of Afghanistan, which was largely based on satellite imagery (see Figure 17.3) because of the danger of field work in the Taliban-infested countryside in a time of war. Even our estimate of the location and magnitude of the 1505 earthquake is based largely on Babur's single account. In the United States, the standard of professional practice to help assess the earthquake hazard to a major city would be to conduct backhoe trenching investigations to identify past earthquakes on the fault, including the 1505 earthquake. We know where the fault is, we just don't know its earthquake history.

The USGS and my consulting firm have submitted proposals to USAID to locate places within and around Kabul that would be safer against earthquakes. A zoning map of the city could lead to building codes that, if enforced, could strengthen apartment buildings against collapse during the inevitable repeat of the 1505 earthquake. We have proposed paleoseismic trenching to learn the history of several mapped faults near Kabul and thereby assess the level of danger. These unsolicited proposals have stimulated little response, and in some cases USAID has not even acknowledged receiving them.

The current American plan is to cease combat operations in hopes that the Afghan government at all levels can take over its own security against the Taliban insurgency. At the same time, pervasive corruption in Kabul and other Afghan cities means that the building boom which is now underway is being done without any consideration of resistance to earthquakes. (Transparency International ranks Afghanistan as one of the most corrupt countries on Earth.) Many returning refugees have simply taken matters into their own hands and are living in slums that would collapse in an earthquake much smaller than magnitude 7.3 (Figure 17.2). USAID supported our efforts to strengthen new buildings in the earthquake-shattered Nahrin region, but that support was after the fact. No funds were allocated to plan for the next earthquake in Afghanistan's largest city.

What would happen if the repeat of the 1505 earthquake occurred in the next few years? A large number of buildings would collapse, including the slums, many of which are constructed of buildings of mud-block construction like those in Nahrin, and of unreinforced concrete. The loss of life could be in the hundreds of thousands, a human catastrophe. Would this outcome cause government at all levels to become nonfunctional, a failed state, with militias and warlords taking over? Have we considered the implications of this outcome even after the American investment of hundreds of billions of dollars in defeating the Taliban insurgency and attempting to build an infrastructure?

18 Earthquakes in the Himalaya

My first view of the High Himalaya came in the 1980s, when I was invited to visit Professor K. S. Valdiya, head of the Geology Department at Naini Tal University, situated in a lovely alpine hill town in the Kumaun Himalaya of India. Professor Valdiya was one of the first geologists to recognize that some of the major faults of the Himalaya are active. Naini Tal is in the mountains, so I had to travel there by road. I took a bus from the Old Delhi bus terminal and was on my way.

I was met by two junior faculty members and taken to the university hostel to meet Professor Valdiya the following day. But when I went to the Department the following morning, I was told that Professor Valdiya was not available. He was with the vice chancellor, which is the Indian equivalent of the president of the University. Hmmm, I thought, that's strange. After inviting me here, he wasn't on hand to greet me at his office. Oh well, when the boss calls, what can you do?

Later in the morning, Professor Valdiya returned to the department, out of breath and highly embarrassed. The vice chancellor had indeed been in his office, but he was trapped there by a group of angry, protesting Kumauni students. The vice chancellor had the misfortune of being a Punjabi from the plains, and thus no more welcome to run a Kumauni university in the mountains than an Englishman would have been decades before. Professor Valdiya is a Kumauni and could communicate with the protesters in their own language. He successfully placated everybody, and the students left. Crisis resolved.

On one of the days I was there, Professor Valdiya took me up a road high above the town to a place called Snow View. We got out of the car and looked north to a wall of snow peaks, the High Himalaya, including a famous mountain, Nanda Devi. The mountain is sacred to many people, some of whom have named their daughters Nanda Devi, including Willi Unsoeld, one of the first Americans to climb Everest. I was in the presence of the greatest mountain range on Earth. The highest mountains of the Himalaya are in such a special class that many of them are rarely called Mount X. One would never say Mount Nanda Devi, just Nanda Devi.

A few years later, in 1991, my son Steve and I made a pilgrimage to Nepal to Everest Base Camp, at 17,598 feet (5364 meters) above sea level, to view the highest point on the planet and to watch cloud banners flaring past the summit of Everest. My wife Jean, Steve's mom, had died of cancer earlier that year, and for us it was a memorial trip. Steve and I built a rock cairn in her memory on another summit, Chukhung, with a commanding view of the south face of Lhotse, the fourth highest mountain on Earth.

My life is forever linked with the Himalaya, which is unlike any other place on Earth. The Himalaya is one of the world's most beautiful places, but it has a dark secret: earthquakes.

TECTONIC ORIGIN

Most mountains are high because they are uplifted as part of plate tectonics, and the Himalaya is uplifted more than any other mountain range on Earth. The reason is that the mountain range is the site of collision between the continent of India, which started its journey more than 70 million years ago thousands of miles to the south, north of Antarctica, and traveled northward to the southern edge of Eurasia. The boundary between Eurasia and India is a zone where plates are converging, but unlike subduction zones on Earth involving one or two oceanic plates, this one pits continent against continent. India is being driven beneath Tibet at a rate of two-thirds of an inch (18–22 mm) per year, with the two continents first coming into contact about 50 million years ago. Remnants of the Indian continent can be detected geophysically at great depth as far north as central Tibet.

Because of the continental collision, Tibet is the highest plateau on Earth. The uplift of both the Himalaya and the Tibetan Plateau is a product of that collision.

This was the reason for my visit to Professor Valdiya at Naini Tal. Kumaun is close to the southern edge of this great collision zone, and Professor Valdiya was a pioneer in the study of the plate-boundary faults between the Himalaya and India, close to his university. He was able to recognize the significance of this boundary, and, as a Kumauni, he took time to give talks to local villagers in their own language, explaining the great tectonic collision in Kumaun and the faults that marked its boundary.

The southern boundary of the High Himalaya is marked by an inactive thrust fault in which the high peaks have been thrust over ancient sedimentary rocks called the Lesser Himalaya (Figure 18.1, bottom diagram). The Lesser Himalaya, in turn, is thrust southward over the Sub-Himalaya, which is made up of sandstone, conglomerate, and shale, erosional products of the uplift that accompanied the collision. The Sub-Himalaya is itself thrust southward over the river deposits of the Ganges Plain along an active fault called the Main Himalayan thrust or, locally, the Himalayan Front thrust (Figure 18.2). This thrust at the base of the Sub-Himalaya is the surface expression of the true plate boundary between the Indian plate and Eurasia.

In the northern part of the Lesser Himalaya, beneath its faulted boundary with the High Himalaya, a band of earthquakes extends along the entire Himalaya from Pakistan eastward to Assam (Figure 18.1, upper diagram). This band of seismicity sometimes includes damaging earthquakes, including the 1991 Uttarkashi earthquake of magnitude 7 and the 1999 Chamoli earthquake of magnitude 6.4, both in the Garhwal Himalaya (Figure 18.1, lower diagram). I visited the region struck by the Chamoli earthquake and saw the damage that it did to small mountain villages, where many people lost their lives, some by landslides shaken loose by the earthquake. This band of seismicity also marks the plate boundary at

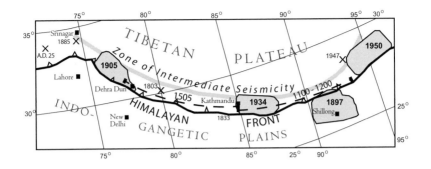

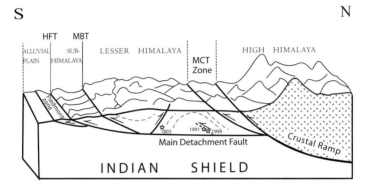

FIGURE 18.1 (Top). Heavy line: Himalayan Front thrust, active plate boundary between the Indo-Gangetic Plains (Indian plate) and Tibetan Plateau (Eurasian plate). Numbers and shading mark four major earthquakes in 1897, 1905, 1934, and 1950; Xs locate additional earthquakes with dates. Gray line marks zone of moderate seismicity, including a large earthquake in Pakistani Kashmir in 2005. Line numbered 1100–1200 and 1505 locate Senthil Kumar's controversial interpretation of two megaquakes larger than the earthquakes above (see Figure 18.3). (Bottom). Block diagram and cross-section of the Garhwal Himalaya in northwest India, showing major faults and the sources of several major earthquakes. HFT, active Himalayan Front thrust fault; MBT, Main Boundary thrust fault; MCT, Main Central thrust fault (inactive). Source: modified from Yeats (2012, Fig. 8.6) and work by V. C. Thakur of Wadia Institute of Himalayan Geology, Dehra Dun.

depth, and the dip of the plate-boundary faults is steep enough that the collision along this fault is accompanied by uplift of the Great Himalaya to the north (Figure 18.1, lower diagram).

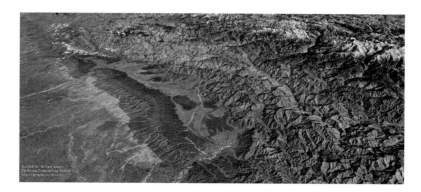

FIGURE 18.2 Oblique digital image of Himalayan front in northwest India, view to northwest. Edge of hills is marked by a plate-boundary thrust fault between the Ganges Plain (left) and Himalayan foothills. Plate boundary is marked by a great reverse fault and by uplift of a fold, behind which is the Dehra Dun Valley, which is itself bounded on its north side by another active fault. The Ganges River flows along the eastern edge of this valley, and the Yamuna River flows along its western edge. Source: image by Dr. William A. Bowen – California Geographical Survey – http://geogdata. csun.edu. Compare with Figure 18.1 block diagram.

But the band of seismicity is many miles from the Himalayan Front thrust, and the intervening foothills have only low instrumental seismicity. Geophysical surveys show that the low hills between the band of seismicity and the Himalayan Front thrust overlie a part of the plate-boundary thrust that is nearly flat lying (Main Detachment fault, Figure 18.1).

The Himalaya, 1500 miles (2500 km) long, is broadly curved in map view, terminating abruptly on west and east against active plate-boundary strike-slip faults. The western boundary of the Indian tectonic plate is the location of the Hindu Kush Mountains of northeastern Afghanistan and the Chaman strike-slip fault of Afghanistan and Pakistan (see Figure 17.3), discussed in the previous chapter. Along the Chaman fault, the Indian tectonic plate grinds northward past a separate block farther west in Afghanistan. The eastern boundary is another strike-slip fault, the Sagaing fault of Myanmar, or Burma, which also records the northward movement of India past

Southeast Asia (see Figure 19.1). At the western and eastern boundaries of the Himalaya, the major rivers of Tibet and Nepal plunge down gorges that are as much as four times the depth of the Grand Canyon. One peak, Kailas, in the central Himalaya of Tibet, is sacred to several of the world's great religions, and for that reason it has never been climbed. The Indus, Brahmaputra, Ganges, and Sutlej rivers all have their headwaters on the slopes of Kailas. Many people, including my son, Steve, have walked around the base of Kailas, as if it were a gigantic prayer wheel. This walk takes about a week.

THE GREAT TRIGONOMETRIC SURVEY

The British were fascinated by the land of India that had come under first the rule of the British East India Company and then, in the mid-nineteenth century, under direct rule of the British Crown. They began to survey their new domain, including the wall of snow peaks to the north, in a campaign called the Great Trigonometric Survey. One of the survey heights, Peak XV, was later named for the former Surveyor General of India, Colonel George Everest. Another mountain in the Karakoram Range in the far western Himalaya was given its survey name, K-2, or Karakoram 2. Despite efforts to re-name it for another British dignitary, it has retained its original survey name, K-2, to the present day. It is the highest mountain in Pakistan and the second highest mountain on Earth. Everest has retained its British name, although it is Sagarmatha in Nepal and Qomolungma in Tibet.

HIMALAYAN EARTHQUAKES

It is clear that the Himalaya, in addition to being a beautiful mountain range, is subject to earthquakes. This was first observed by Thomas Oldham, who in 1850 was appointed the first director of the Geological Survey of India. His son, Richard, used Thomas' notes to publish an earthquake catalog in 1883. A huge earthquake struck northeastern India in 1897 and was studied in detail by Richard Oldham. Among Richard's observations was that objects on the surface were thrown into the air during the earthquake, indicating that the upward acceleration of the ground due to the earthquake was greater than the downward acceleration due to gravity.

Led by the work of the Geological Survey of India, it was concluded by the middle of the twentieth century that in the past 100 years, the far north of India had been struck by four earthquakes with magnitudes greater than 8 (Figure 18.1, upper diagram). These were the northeast India earthquake of June 12, 1897, the Kangra earthquake of April 4, 1905 in northwest India, the Nepal–Bihar earthquake of January 15, 1934 south of Kathmandu, and the Assam earthquake of August 15, 1950 in easternmost India and adjacent China.

But there was a problem with all of these earthquakes. Despite their great size, few reports of surface rupture had been documented, although Oldham did report surface rupture in 1897 of a secondary fault in the Shillong Plateau south of the Himalaya. In response, different groups of scientists began to investigate each of these earthquakes in greater detail, using technology unavailable to the Oldhams and their immediate successors.

The Kangra earthquake of 1905, in which 20,000 people were killed, was re-examined by Roger Bilham and his students at the University of Colorado, who found that there were actually two earthquakes, one in the Kangra, Dharamsala and McLeodganj region and another, smaller one farther east, close to Dehra Dun. (Long after the earthquake, McLeodganj would become the center of exile of the exiled Dalai Lama and his followers.) The magnitude was demoted from greater than 8 to 7.8, still a very strong earthquake, but much smaller than Richard Oldham had thought. The main shock was close to Kangra, within the zone of high seismicity close to the foot of the Great Himalaya, but the earthquake rupture then propagated south of the epicenter at depth along the flat part of the plate-boundary thrust. However, it did not propagate all the way to the Himalayan Front thrust, and so there was no rupture at the surface.

The 1934 Nepal–Bihar earthquake was studied by two groups, one from the Geological Survey of India and another centered in Nepal. These surveys were done independently of each other because

Nepal at the time of the earthquake was closed to foreigners, and there was essentially no interaction between the two groups. The magnitude was estimated as high as 8.4, and as many as 30,000 people lost their lives. One view, based on extensive ground deformation in India, was that the earthquake rupture extended south of the Himalayan front, although much later this area, called the "slump belt," was reinterpreted as a result of strong shaking in the water-saturated floodplains of northern India. In Nepal, the earthquake was called the Great Kathmandu Earthquake because of the heavy damage to Nepal's capital city. The epicenter was south of Everest, and damage was heavy at the famed Thyangboche Monastery. More recently, teams led by Soma Sapkota of the Nepal Department of Mines and Geology, together with others at the Earth Observatory Singapore and the Institut de Physique du Globe de Paris, found evidence of surface rupture related to the 1934 earthquake. Its magnitude is now estimated as 8.1.

The 1897 earthquake had been described by Richard Oldham as the Assam earthquake in the eastern Himalaya, with the loss of 1500 lives. It was studied in detail by Roger Bilham and his colleague Philip England of Oxford University and C. P. and Kusala Rajendran, a husband-and-wife team now at the Indian Institute of Science at Bangalore, with the result that the source fault of this earthquake is now located not in the Himalaya but at the northern edge of the Shillong Plateau, south of the Himalaya and south of the Brahmaputra River. Its magnitude was revised to 8.1. Bilham was able to re-examine nineteenth-century records of the Great Trigonometric Survey preserved in England and confirm that a fault at the northern edge of the Shillong Plateau was the source of the earthquake. Bilham and his colleagues named this surface fault the Oldham fault after Richard Oldham. Their work established that the Shillong Plateau was itself a major earthquake source, with faults on both its northern and southern margins.

A fault on the south side of the Shillong Plateau, called the Dauki fault, is a major hazard to the city of Dhaka, capital of

Bangladesh, an earthquake time bomb with a population close to seven million. Bangladesh and the adjacent Indian state of Bengal are inhabited by Bengalis, who are Muslim in Bangladesh and Hindu and other religions in the State of Bengal. Bangladesh has essentially no effective earthquake-resistant building codes, and so a major earthquake on either the Dauki fault or the Oldham fault will be a catastrophe. Much of Bangladesh is built on soft sediments of the deltas of the Ganges and Brahmaputra rivers, which are subjected to strong shaking (bowl of jello effect, as discussed for Mexico City in Chapter 25). This problem, plus major corruption in the building construction industry, has set the stage for losses of life in the hundreds of thousands in the next earthquake. Dhaka has developed a major industry in clothing manufacture for brands in the West, including Walmart and Gap, but some of the clothing factories have collapsed without an earthquake, accompanied by great loss of life.

The 1950 earthquake is the least known of the four great earthquakes, in part because the zone of great damage is on both sides of the border with China, where it is called the Chayu earthquake. It was assigned a magnitude of 8.4, making it the largest of the four earthquakes and the largest in Chinese history recorded on seismographs. Like the other three, it had no evidence of surface rupture, but in part this could be due to its remoteness, partly in the jungles of Assam close to a border between two countries with relatively poor relations. Because of the remoteness of the damage zone and its low population, the number of deaths was relatively small.

Even with better knowledge, the source-rupture problem remained. None of the four earthquakes had large surface displacement, and at least one, the Kangra earthquake, had no surface rupture at all. However, a relatively well-known exposure of the Himalayan Front thrust near the village of Mohand, south of Dehra Dun (edge of the mountains as shown in Figure 18.2), clearly shows the deformed sedimentary deposits of the Sub-Himalaya thrust over flat-lying sediments of the Ganges Plain, sediments containing archaeological artifacts, including pottery. What earthquake formed the Himalayan Front thrust?

As part of a workshop in 1996 on active faults in the Himalaya, sponsored by the Indian Department of Science and Technology and the American National Science Foundation at the Wadia Institute of Himalayan Geology at Dehra Dun, an Oregon State University graduate student, Emily Oatney, and an Indian scientist from Wadia, N. S. Virdi, excavated a paleoseismic trench at a fault close to the Giri River in the western end of the intermontane valley including Dehra Dun. (This fault had been first described by Richard Oldham in 1881 while he was in the Himalaya recovering from malaria.) Oatney and Virdi were unsuccessful in trenching across the fault scarp, but they did trench two wedges of gravel shed from the fault scarp during an earthquake. They found enough radiocarbon to date the gravel wedges and thereby date the earthquakes that generated them. They were surprised to find that both wedges were formed between the middle of the fifteenth century to the middle of the seventeenth century, much younger than she and Virdi had expected.

In the field, I met a young Indian geologist, Senthil Kumar, who planned to pursue a PhD at the University of Nevada Reno under the supervision of my colleague and friend Professor Steve Wesnousky. Kumar's thesis project was to trench the Himalayan Front thrust in as many places as he could and to date the most recent earthquake on the thrust. Five of the six trenches he excavated contained information on the age of the most recent fault event in the fifteenth century, consistent in timing with Oatney and Virdi's results, except Kumar's dates were on the Himalayan Front thrust itself. Although no earthquake had been reported at the exact time of Kumar's dates, an Indian scientist, R. N. Iyengar, had described a large historical earthquake on June 6, 1505 (the same year as a large earthquake on the Chaman fault near Kabul, discussed in the previous chapter). Kumar suggested that this earthquake might be the same earthquake as those recorded in his trenches, if some of the carbon he dated had been reworked. Kumar concluded that surface rupture accompanying the 1505 earthquake might have extended from a trench he excavated near Chandigarh, a planned city that is the capital of two Indian states, Haryana and Punjab, east 540 miles (900 km) to the Himalayan front south of Kathmandu, an interpretation shown in Figure 18.1. Surface rupture accompanying the most recent earthquake on the Himalayan Front thrust was locally greater than 60 feet (16–26 m), the largest on any continental thrust fault on Earth. This suggested that the 1505 earthquake was larger than any of the four that had been described a half-century earlier and could explain the most recent large-scale surface rupture on the Himalayan Front fault.

South of Kathmandu, a team led by a Nepali geologist, B. N. Upreti, and Professor Takashi Nakata of Hiroshima University, found trenching evidence for a surface rupture on the nearly flat-lying Himalayan Front thrust dated as about 1000 years ago. A surface rupture on the Himalayan Front thrust farther west in Nepal was dated as about 1100 years ago, with a surface rupture of at least 40 feet (12–17 m). The length of the surface rupture based on trenching is at least 500 miles (800 km), similar to the younger earthquake to the west (Figure 18.1). However, as with the younger earthquake, there is not a close correlation in ages based on paleoseismic trenches and historical ages. Kathmandu was destroyed in an earthquake in AD 1255, in which the king, Abhaya Malla, was killed. However, this age is slightly younger than the ages from trenches. It seems unlikely that the 1255 earthquake would not have been recorded in trenches, or that the older ages in the trenches document an earthquake missing in the historical record.

In a book I published in 2012, *Active Faults of the World*, I interpreted these two medieval superquakes as having their epicenters on the zone of active seismicity, but instead of stopping just to the south, as the 1905 Kangra earthquake did, continuing across the flat thrust of the Lesser Himalaya and Sub-Himalaya to the Himalayan Front thrust, thereby generating two huge earthquakes. The recurrence interval of these superquakes would be 1000–3000 years.

I followed Senthil Kumar's interpretation in concluding that these two superquakes, not the four that were thought to be the largest a half-century ago, should be the ones considered as the maximum-size earthquake that could strike the region and affect the stability of major dams and power plants in addition to the safety of the millions of people who live nearby.

However, the lack of close correlation of contemporary evidence for earthquakes with the ages of faulting based on paleoseismic trenches led scientists, including J.-L. Mugnier of the University of Savoie in France, R. Jayangondaperumal of the Wadia Institute of Himalayan Geology, and B. Upreti of Tribhuvan University in Kathmandu, to re-examine new paleoseismic and historical evidence for earthquakes over the last 1000 years. This evidence led them to conclude that the interpretation of two superquakes was oversimplified and not supported by paleoseismic evidence from an increasing number of trenches, as well as a reinterpretation of historical evidence (Figure 18.3, lower diagram).

They conclude that earlier earthquakes could have been larger than the four earthquakes described by Richard Oldham, but not as large as had been thought by Kumar. They proposed a westward-propagating sequence of earthquakes in AD 1100, 1255, 1344, and 1430, with the well-known 1505 earthquake east of the 1430 earthquake (Figure 18.3). The largest earthquake in their interpretation was the 1344 Kathmandu earthquake, with a magnitude of 8.6, but even larger earthquakes are possible. However, additional paleoseismic trenching will probably modify this interpretation.

On April 25, 2015, an earthquake on the plate-boundary thrust of magnitude 7.8, with its epicenter in the Gorkha region northwest of Kathmandu, directed its seismic energy eastward, causing major destruction and loss of life in Kathmandu, nearby Bhaktapur (Figure 18.4), and isolated mountain villages, and as far east as Everest Base Camp, where climbers and their Sherpa guides in the Khumbu Icefall preparing to ascend Earth's highest mountain were struck by an avalanche triggered by the earthquake.

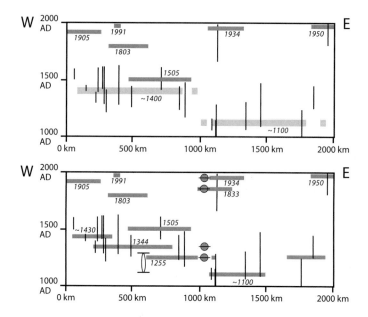

FIGURE 18.3 Two interpretations of pre-instrumental earthquake history of the Himalaya. Horizontal shaded lines show dimensions of historical earthquakes discussed by Richard Oldham and others at the Geological Survey of India, plus historical earthquakes in 1505 and 1803 and the Uttarkashi earthquake in Garhwal. The 1897 earthquake in Shillong Plateau is not included. Vertical lines are age limits of earthquakes identified in paleoseismic trenches. Light shading shows dimensions of the 1400 and 1505 superquakes of Senthil Kumar. Lower diagram: Interpretation of Mugnier *et al.* based on more complete paleoseismic and historical evidence, showing more and larger earthquakes than those identified by Oldham but not as large as interpreted by Kumar. Note the absence of earthquakes west of the Kathmandu (Nepal–Bihar) earthquake of 1934, a gap partially filled by the 2015 Gorkha earthquake of magnitude 7.8 northwest of Kathmandu. Shaded circles: earthquakes damaging Kathmandu. Source: Mugnier *et al.* (2013), used with permission.

The earthquake struck within a seismic gap identified by the late K. N. Khattri of Roorkee University in India between the 1934 Kathmandu earthquake and the 1505 earthquake to the west in northern India. Like the 1991 Uttarkashi and 1999 Chamoli earthquakes (block diagram, Figure 18.1), the Gorkha earthquake struck

FIGURE 18.4 Soldiers with buildings damaged by the April 12, 2015 earthquake in Bhaktapur city, 20 km from the Nepali capital, Kathmandu. Source: image (taken April 30, 2015) courtesy of think4photop / Shutterstock.com.

close to the ramp beneath the High Himalaya and the flat thrust farther south in the direction of the Himalayan front.

On May 12, another earthquake of magnitude 7.3 struck the region between Kathmandu and Everest Base Camp, causing additional loss of life and heavy damage, not only to historic World Heritage Site buildings in Kathmandu, but also to isolated mountain villages to the east, including Langtang Valley, a scenic trekking route visible from Kathmandu. The location of the earthquake was on the same fault that ruptured in April.

This section closes with an earthquake in that part of Kashmir under the control of Pakistan, called Azad Kashmir (Free Kashmir). The Kashmir earthquake, with its epicenter close to Muzaffarabad, the administrative center of Azad Kashmir, struck on October 8, 2005 with a magnitude of 7.6. The official death toll in Pakistan was 75,000, with another 1400 killed in adjacent Indian-administered

Kashmir. The earthquake, like the subsequent earthquakes close to Kathmandu, struck within the belt of high seismicity that extends across the Himalaya north of the Himalayan range front (Figure 18.1 map and block diagram), although it was larger than other recent earthquakes in this zone, including the Uttarkashi and Chamoli earthquakes. Much of the region of strongest shaking was in the mountains, and landslides accounted for many of the deaths. Rescue operations were spearheaded by helicopter units from both the United States and Great Britain, thereby saving many lives. The Kashmir earthquake was large enough to be accompanied by surface rupture on a fault with moderately steep dip.

POPULATION AT RISK

Unlike many of the other sections of this book, this section is centered on a seismically active region, the Himalaya, rather than a single metropolitan area. The reason for this is that much of the population increase near the Himalaya is in the rural districts, in addition to major cities. For example, the death toll of the 2005 Kashmir earthquake included many people in the back country, who had to be rescued by helicopter since the roads were largely blocked by rockfall and landslides. The Pakistan government tried to persuade the people to be evacuated to refugee camps in the lowlands, but this would have required them to leave their animals behind, which they refused to do. Farm animals to a family in the Himalaya are the counterpart of a bank account in the United States, and they would no more leave their animals behind than we would leave our money behind. This was not initially appreciated by the Pakistan government, dominated by people from the plains south of the mountains.

The largest city in the earthquake area is Muzaffarabad, with a population of 725,000. When we arrived there in early 2006 to map the surface rupture, we were greeted by a large group of people hoping we could work out the origin of the earthquake and prevent another earthquake from striking them. We excavated a paleoseismic trench on the earthquake fault, naming it the Jasmine Memorial Trench and dedicating it to the wife of a judge who owned the property and did all he could to help us. Jasmine had been killed in the earthquake, and tears were sometimes shed when we met Judge Gillani for tea. I still hear from him.

Several cities in and near the Himalaya have large populations. Peshawar, the capital of the Pakhtunkhwa province in Pakistan, containing most of Pakistan's Pashtun population, has a population of more than 3.5 million. The twin cities of Islamabad and Rawalpindi have a combined population of more than 2.3 million. Srinagar, in Indian-administered Kashmir, has a population of 1.2 million. Kathmandu, with a population of 1.5 million in the Kathmandu Valley, was heavily damaged by earthquakes in 1255 and 1344 as well as by the 1934 Nepal–Bihar earthquake and the two earthquakes in 2015 (Kathmandu is located by the circles in lower diagram of Figure 18.3). Some of the hill towns that were refuges from the heat of summer during the days of the British Raj have grown to large populations, including Shimla, at more than 813,000, and Dehra Dun at 560,000.

The problems are similar to those in other places in the developing world that are at risk from earthquakes. The governments attempt to deal with the situation after the fact, but in many cases aid from the United Nations and the developed world is inefficiently used, or is siphoned off by corrupt officials. On the other hand, working groups have been organized in Myanmar and Kathmandu to strengthen against earthquakes, as discussed in Chapter 19.

With assistance from a nonprofit consulting group called Geo-Hazards International (GHI), formed in 1991 by an American seismologist, Dr. Brian Tucker, the Kathmandu Valley has created a Disaster Management Unit as part of the city government. The disaster management plan has as its objective raising awareness of earthquake risk, especially in public schools. With guidance from GHI and from Nepali scientist Amod Dixit, Kathmandu has strengthened a non-governmental organization, the National Society for Earthquake Technology – Nepal, to establish committees of citizens within local government to raise awareness and upgrade structures. Nepali masons trained to construct earthquake-resistant buildings in the Kathmandu Valley traveled to Gujarat, India, heavily damaged in an earthquake in 2001, to train local masons there in earthquake-resistant structures. In recognition of his leadership, in 2001, the King of Nepal awarded Dr. Tucker the Gorakha Dakshin Bahu medal for service to the people of Nepal. He was also named a MacArthur Fellow in 2002, receiving an award sometimes called the Genius Grant, awarded by the John D. and Catherine T. MacArthur Foundation to highly creative people who are committed to "a more just, verdant, and peaceful world."

After the Kashmir earthquake of 2005, GHI, with support from USAID and the Pakistan Higher Education Commission, teamed with engineers at the NED University of Engineering and Technology (www.neduet@edu.pk) in Karachi, Pakistan's largest city, to start a three-year project to strengthen the curriculum in retrofitting unsafe buildings, including the commonly built reinforced concrete buildings with unreinforced infill walls. In addition, the project includes workshops to involve practicing professionals as well as students, with the aim of using the 2005 earthquake to motivate builders in Pakistan and elsewhere to make their buildings earthquake-resistant. For details, go to http://peer.berkeley.edu/pdf/Khalid-pakistan-final2.pdf.

GHI is also working on strengthening hospitals and schools in Bhutan and India with funding from the Global Facility for Disaster Reduction and Recovery (GFDRR). Bhutan was damaged by earthquakes in 2009 and 2011. India is operating under a National Action Plan focused on hospitals and schools through their National Disaster Management Authority. Support is also provided by the Bechtel Foundation and Klehn Family Foundation.

19 Myanmar and the Sagaing fault

Burma, renamed Myanmar by its government in 1989, is one of the world's poorest countries, even though it is rich in natural resources. It has a written history lasting more than 2000 years, and the country has been predominantly Buddhist for most of that time. It was a major rice-exporting country before World War II. The kingdom fell to British rule in the nineteenth century, although the British occupation was marked by riots, rebellion, and demonstrations, including protests by Buddhist monks. A strike in 1936 at Rangoon University led by a student leader named Aung San was put down violently.

In the spring of 1942, the country was overrun by the Imperial Japanese Army, and Allied actions against the Japanese in Burma used the famous Burma Road from southwest China. In 1945, the Japanese were expelled; Aung San was one of the leaders advocating independence from British rule. On July 19, 1947, Aung San was assassinated, and the anniversary of that day is now celebrated as Martyr's Day. Independence was achieved the following January, although Burma elected not to join the British Commonwealth. The early days of independence were marked by attempts at revolution, some led by Communists.

In 1962, the government of Burma was overthrown in a military coup, and a half century later, the military is still in charge. After 1962, separatist movements arose, particularly in the Shan States in the eastern part of the country. Some of these separatist movements are still active, including Muslim riots in the western part of the country close to Bangladesh. After a mass uprising in 1988 supporting democracy, the governing junta permitted elections to be held in May 1990, with the National League for Democracy, led by Aung San's daughter, Aung San Suu Kyi, winner in a landslide victory.

However, the military government refused to abide by the results of that election, and Aung San Suu Kyi, winner of the Nobel Peace Prize in 1991, was kept under house arrest until the regime allowed democratic reforms in 2011–2012. She was elected to Parliament in 2012, and the regime appears to be emerging from its isolation and sanctions as a pariah state, although demonstrations against the Muslim minority continue.

POPULATION EXPLOSION AND THE SAGAING FAULT

Many of the demonstrations by students and Buddhist monks had been centered in Rangoon, renamed Yangon, and the government responded in 2005 by moving the national capital 200 miles to the north, to a rural area between Yangon and Mandalay, Myanmar's second largest city. The new city was named Naypyidaw. The relocation of government ministries and military bases has resulted in Naypyidaw's becoming Myanmar's third largest city, and one of the ten fastest growing cities on Earth.

Much of the population and culture are found in the Central Basin between the Indoburma Ranges, also called the Arakan or Rakhine Yoma, on the west, and the Shan Plateau on the east (Figure 19.1). The three largest cities are Yangon, population 5.4 million, Mandalay, population one million, and Naypyidaw, population 900,000, a total of 7.3 million in those three cities alone. Population in all three cities has grown rapidly, in part because of a high birth rate and because of the immigration of large numbers of Chinese from across Myanmar's northeastern border with China. Yangon's population in 1930, the date of the last major earthquake there, was fewer than 200,000. Mandalay had a population of 250,000 as recently as 1960, and, of course, Naypyidaw, the new capital, did not exist at all until 2005.

The Central Basin contains the Sagaing active fault, which is the tectonic plate boundary between the Sunda plate on the east and the Burma plate on the west, which is transitional to the Indian plate farther west (Figure 19.1). The cities of Yangon, Naypyidaw, and

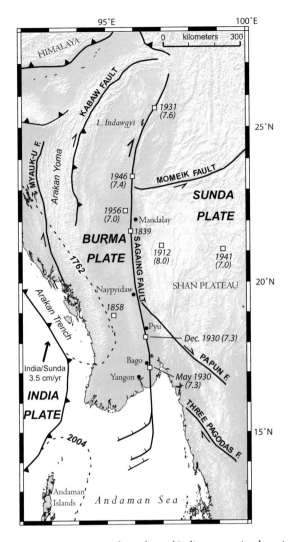

FIGURE 19.1 Eastern boundary of Indian tectonic plate, including Sagaing
strike-slip fault, which threatens Myanmar's two principal cities, Yangon
and Mandalay, and its new capital, Naypyidaw. Open squares locate
major earthquakes with year of occurrence and magnitude. Dashed lines
mark the area of strong ground motion of the northern end of the
Sumatra–Andaman earthquake of 2004 and of a subduction zone
earthquake in 1762. The Bago earthquake of magnitude 7.3 struck close
to present-day Yangon, then the small city of Rangoon, in May 1930.
Source: modified from Yeats (2012, Fig. 8.4) and Tsutsumi and Sato (2009).

Mandalay are all close to the fault and therefore all subject to earthquakes. The Sagaing fault has a slip rate of about two-thirds of an inch (18 mm) per year, which appears to be very slow, but it is high enough to have been the source of at least six damaging earthquakes during Burma's long written history (Figure 19.1). Mandalay was struck by earthquakes in 1839 and 1956, each one of magnitude 7.

Rangoon was heavily damaged by an earthquake of magnitude 7.3 on May 5, 1930, destroying the town of Pegu (present-day Bago), east of Rangoon, and killing more than 500 people, mainly in Pegu. More than 50 people were killed in Rangoon, more than 50 miles (80 km) away from the epicenter. The earthquake was accompanied by displacements larger than 10 feet (3 m) along the Sagaing fault, as determined by a Japanese team led by Hiroyuki Tsutsumi of Kyoto University. The relatively small number of deaths was due to the low population, fewer than 200,000 in Rangoon, and the fact that many people lived in timber houses that were less likely to collapse. Today, the population of Yangon, as Rangoon is now called, is more than 20 times larger, with much of the population living in poorly constructed brick buildings built after 1990, indicating that the next large earthquake to strike Rangoon will be accompanied by heavy losses of life.

The Sagaing fault is a major part of the boundary between the Indian subcontinent and the Shan Plateau. The Indian plate is driving northward against Tibet and forcing up the Himalaya, the highest mountain range on Earth. However, the strike-slip rate between the Indian and Sunda plates is twice as fast as the slip rate on the Sagaing fault alone; the plate rate is close to an inch and a half (35 mm) per year. The Burma plate is caught in a vise between the Indian and Sunda plates, and some of the slip is taken up on other active faults to the west, closer to the Bay of Bengal.

SUBDUCTION ZONE EARTHQUAKE IN 1762

One of these faults taking up the additional displacement may be a subduction-zone fault causing tectonic transport of oceanic crust in the Bay of Bengal beneath western Myanmar (Figure 19.1). West of the Andaman Islands and west of the Indonesian island of Sumatra, this fault is expressed at the surface as the Sumatra trench, and the plate boundary has also been the source of active volcanoes on Sumatra. On December 26, 2004, this fault broke in a superquake of magnitude 9.15 that produced a tsunami that swept over the northern tip of Sumatra and came ashore on the west coast of Thailand and the east coast of countries west of Myanmar, including Sri Lanka, southernmost India, and eastern Africa. More than 300,000 people lost their lives, with the largest number on the northern end of Sumatra.

The northern end of the 2004 earthquake was in the Bay of Bengal (Figure 19.1), south of the Irrawaddy (Ayeyarwady) Delta of Myanmar, but the fault itself may have continued northward into Myanmar. The evidence for this includes young volcanoes in the Central Basin of Myanmar and deep earthquakes marking the zone of subduction. This area of Myanmar was the site of a very large earthquake on April 2, 1762, called the Rakhine or Arakan Yoma earthquake (Figure 19.1). This earthquake uplifted the coastline, including offshore islands. Dated coral remnants of marine shorelines along the Burma coast document stair-stepped marine terraces of three different ages, each caused by uplift in a subduction zone earthquake. The uplifted terraces indicate an earthquake recurrence interval on the subduction zone of 1000 to 2000 years.

Some scientists suggested that the 1762 earthquake produced a major tsunami, which, if repeated today, would result in huge losses of life because of the millions of people in Myanmar, Bangladesh, and eastern India that live close to sea level on the coast of the Bay of Bengal, particularly those living in the broad delta of the combined Ganges and Brahmaputra rivers. However, an Indian research group led by Professor Harsh Gupta concluded that this earthquake did not generate a tsunami because the source fault reaches the surface too close to shore. The Indian archives yielded no evidence for a tsunami in 1762.

It does not appear that the southern end of the rupture zone of the 1762 earthquake adjoins the northern end of the rupture zone of the 2004 superquake, and the area between the two earthquakes may be in a seismic gap (Figure 19.1). This gap includes the Irrawaddy delta and the city of Yangon, with its 5.4 million people. This gap lacks evidence for major earthquakes, volcanoes, and uplifted marine terraces, and displacement on earthquakes in this region is probably by strike-slip, not subduction as is the case to the north and south. An earthquake on this section of the subduction zone fault could kill hundreds of thousands of people, but unfortunately the fault is not well mapped in southern Myanmar.

IS THE GOVERNMENT READY FOR THE NEXT BIG EARTHQUAKE?

How well would the government of Myanmar/Burma be able to respond to a very large earthquake striking Yangon, either from the subduction zone or from the Sagaing fault? The government's response to Cyclone Nargis in May 2008 addresses this question. This cyclone, with maximum wind speeds of 135 miles (215 km) per hour, struck the densely populated rice-growing Irrawaddy (Ayeyarwady) Delta. A storm surge as high as 12 feet (3.7 m) acted like a tsunami, drowning many people in the delta. More than 138,000 people lost their lives, and the death toll might have been underreported by the government for political reasons. The cyclone struck during a time when the regime was particularly unresponsive to outside pressure, and airplane relief shipments by the United Nations and individual countries, including the United States, were delayed or denied entry, increasing the loss of life. One would hope that with democratic reforms and the renewed contacts of Myanmar with the outside world in the past few years, aid would be welcomed after the next natural disaster, but the question of how well disaster relief could be delivered remains unanswered.

MYANMAR EARTHQUAKE COMMITTEE

One encouraging sign is the response of local Myanmar scientists and engineers. In 1999, a group of professionals formed the Myanmar Earthquake Committee (MEC), with support from the Myanmar Engineering Society. The MEC has compiled information on faulting, seismicity, and earthquake damage that has been turned over to the government. Myanmar has a building code that is designed especially for local conditions, including shaking response by the beautiful pagodas that dot the landscape in many parts of the Central Basin. One member of the MEC, Tint Lwin Swe, constructed a seismic zoning map of Yangon while he was at Yangon Technical University, and in 2012 the MEC produced a probabilistic seismic hazard map of Myanmar. It is unclear how much effect the MEC has had in avoiding corruption in construction and building inspection to accommodate the rapidly growing population of Myanmar's cities. The government is aware of the earthquake problem and sponsors a seismic network run by the Department of Meteorology and Hydrology. However, the Transparency International Corruption Perception Index ranking of 15 for Myanmar, one of the worst in the world, is not encouraging.

The MEC responded to the September 21, 2003 Taungdwingyi earthquake of magnitude 6.6, including reports of damage to buildings that are unreinforced against strong shaking. This earthquake was about 30 miles (48 km) from Naypyidaw and resulted in 11 deaths. The earthquake might have been caused by a rupture on the Sagaing fault or on a previously unknown fault farther west.

Another significant achievement is the involvement of foreign earthquake scientists in studies of both the Sagaing fault and the Rakhine coast of Myanmar, site of the 1762 earthquake. This collaboration included the involvement of the Japanese team led by Professor Tsutsumi, which studied the May 1930 Pegu earthquake on the Sagaing fault east of Yangon. The MEC, together with the Earth Observatory of Singapore, led by Professor Kerry Sieh, has trenched the Sagaing fault north of Bago, the site of the earthquake in December 1930, discussed above. Professor Kenji Satake, an internationally renowned tsunami expert at the University of Tokyo, has been studying the evidence for tsunamis along the coast of Myanmar. Finally, the Earth Observatory of Singapore is studying the coastal deposits that were deformed in the 1762 Rakhine earthquake.

Starting in 2007, the MEC has organized alternate-year international conferences on earthquake science, with the most recent conference in May 2013, on the fifth anniversary of the Nargis typhoon.

20 Metro Manila, the Philippines

A VIOLENT PATHWAY TO INDEPENDENCE

In the sixteenth century, Spain was a superpower. The King and Queen of León and Castile on the Iberian peninsula had sponsored Columbus' discovery of the Western Hemisphere in four separate expeditions. These were followed by the first voyage around the world by Ferdinand Magellan in 1519–1522, sponsored by King Charles I of Spain, in search of the Spice Islands of present-day eastern Indonesia. The two largest empires in the Western Hemisphere were conquered: the Aztec empire of Mexico by Hernán Cortés in 1521 and the Inca empire of the central Andes of South America by Francisco Pizarro in 1532 (see Chapters 21 and 25 on Peru and Mexico). Both empires yielded vast treasures that were put on ships and sent to Spain. King Charles was later crowned as Charles V, Holy Roman Emperor, the most powerful monarch in Western Europe.

Magellan had claimed the Philippines for Spain during his expedition around the world. He had discovered islands southeast of the main island of Luzon, but he became involved in warfare among individual chieftains, and as a result he was killed in 1521. The rest of the expedition left the Philippines and returned to Europe, in the process completing Magellan's round-the-world voyage. Several expeditions followed, and on one of them the islands were named Islas Filipinas after King Philip of Spain. Philip ordered another expedition to the Spice Islands, southeast of the Philippines, although its real objective was the conquest of the Philippines. A trade route between the Philippines and Acapulco, Mexico and across the Caribbean and Atlantic Ocean to Spain (the Galleon Trade) was initiated, which continued for several centuries and brought great wealth to the Philippines and their colonial masters.

In 1570, a Spanish naval squadron conquered the Kingdom of Maynila, and in 1596, the Governor of the Philippines made Maynila, now renamed Manila, the capital. It has continued to be the capital to the present day. The Philippines were ruled by a Governor General, answering to the King of Spain. Most of the islanders were converted to Christianity except for the Muslim Moros of the southernmost islands west of Mindanao. The Philippines were under Spanish rule for the next three centuries, interrupted only by a brief occupation of Manila by the English in 1762–1764.

In the nineteenth century, Spanish rule became a burden as nationalism spread among the Filipinos. With the help of the United States, looking for a military base in southeast Asia, the Spanish provincial government was overthrown in the Battle of Manila Bay. Spain surrendered not to the native Filipino rebels but to the United States, but the Americans refused to grant immediate independence to the Filipinos. This led to a war between the United States and Filipino insurgents, ending in an armistice in 1902. The policy of the United States was to eventually grant independence to the country, but World War II and invasion by Japan put those plans on hold. General MacArthur was driven out of the Philippines in 1942, but he and his army returned in 1944, and the Japanese were defeated in 1945. The Americans honored their promise of independence, and the Republic of the Philippines was established on July 4, 1946, with Manila as its capital.

GROWTH OF METRO MANILA

Manila has one of the finest harbors in Southeast Asia (Figure 20.1), and it is bordered on the east by a large lake, Laguna de Bay. The Pasig River flows from Laguna de Bay westward to Manila Harbor, dividing the metropolis into two halves. The urban region consists of 13 separate cities and four municipalities; the largest of these is Quezon City. These cities have joined together as a megacity called Metropolitan Manila, which contains 13% of the population of the entire country. Metro Manila's economy generates one-third of the gross national

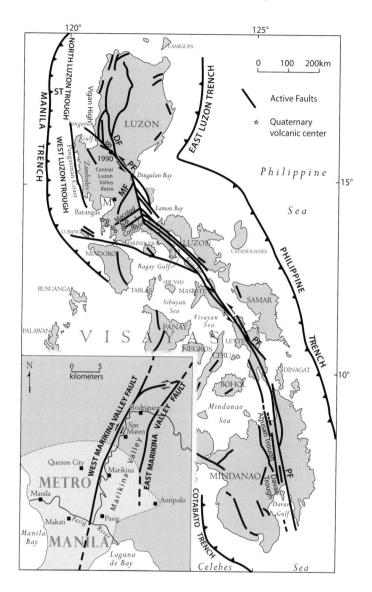

FIGURE 20.1 Active fault map of Philippines. Subduction zones marked by heavy solid lines, with teeth pointed toward the upper plate; crustal faults marked by heavy solid lines without teeth. PF, Philippine fault (strike-slip); DF, Digdig fault that sustained an earthquake of magnitude 7.7 in 1990; ST, offshore Stewart Bank. M, Manila. MF, Marikina Valley faults (see inset). Source: modified from Yeats (2012, Fig. 10.7) and Bischke *et al.* (1990), used with permission.

product of the Philippines. The population has nearly doubled in the last 50 years from six million in 1960 to nearly 11 million in 2010. Metro Manila may be the 28th wealthiest urban center on Earth. Despite its wealth, it contains large slums, and thus it fits the image of a huge city in the developing world, in danger from a large, devastating earthquake.

EARTHQUAKE HAZARD IN THE PHILIPPINES

The Philippine Islands lie within one of the most complex tectonic environments on Earth (Figure 20.1), and the country is at risk from earthquakes, tsunamis, and volcanic eruptions. In recognition of these hazards, the Philippine government formed the Philippine Institute of Volcanology and Seismology (PHIVOLCS) under its American-educated director, the late Ray Punongbayan. Several young Filipino scientists have obtained advanced degrees at universities in Japan, the United States, and elsewhere, and much of the earthquake research in the Philippines is done by these scientists.

The Manila trench marks a subduction zone west of Luzon in which the Sunda plate is being driven beneath the island of Luzon. This trench extends south from Taiwan and west of Luzon and ends west of the island of Mindoro. This subduction zone is deep enough that it has generated volcanoes on islands north of Luzon and on Luzon itself. These volcanoes include Mount Pinatubo, which, in June, 1991, ejected so much material into the stratosphere that it temporarily cooled the climate of Earth, and Taal Volcano, located in Lake Taal, south of Manila, which has erupted 33 times since 1572. Both have erupted in the past century. East of the Philippines, the oceanic Philippine Sea plate is being forced beneath the Philippines along the Philippine trench, which contains one of the two greatest ocean depths on Earth, nearly seven miles (11 km). Both of these subduction zones have generated damaging earthquakes and are capable of earthquakes of magnitude 8 or larger.

In addition to the two subduction zones, the Philippines are crossed by a major crustal fault, the Philippine fault (PF, Figure 20.1), which was the source of the Digdig earthquake of magnitude 7.7 on July 16, 1990, in which more than 1500 people were killed (Figure 20.1). The Philippine fault extends from northern Luzon southeast across the large island of Mindanao (Figure 20.1), with a slip rate of 1–1.5 inches (20–40 mm) per year, released in most parts of the country by large earthquakes. The fault moves by strike-slip; the crust northeast of the fault moves northwest relative to the crust on the southwest side. The fault is roughly parallel to the Philippine trench and is probably related to it. The fault is as close as 60 miles (100 km) from Manila in Dingalan Bay, indicating that it is a major hazard to Manila, leading to damage to the city in the past.

The loss of life in the Digdig earthquake was unusually low for an earthquake of its magnitude. The reason was that most of the houses in the countryside were built simply, and many were simply grass huts (Figure 20.2). Professor Takashi Nakata of Hiroshima University was fortunate to photograph houses along the Philippine fault both before and after the 1990 earthquake, including the grass hut in Figure 20.2. The earthquake knocked down the hut, but its owner simply set it up again after the earthquake; there was no damage and no one was killed. In contrast, the modern Hyatt Terraces Hotel in nearby Baguio was completely destroyed, with many deaths (Figure 20.2).

EARTHQUAKE HAZARD TO MANILA

Ten earthquakes have affected Manila since its founding, and at least six of these did major damage. The first, with a magnitude estimated as 7.5, struck on November 30, 1645, destroying Manila, then a small town, including its cathedral that had been built 31 years before. The Spanish had no idea that Manila was under threat of earthquakes, and as a result the stone cathedral was not reinforced against shaking. The Spanish estimated the loss of life to be 600 people, but they only counted Spanish deaths, not deaths of native Filipinos. Native houses were built of wood with roofs of palm fronds, and these caught fire, killing many thousands of Filipinos. The source of the earthquake was

FIGURE 20.2 Damage from the 1990 Digdig earthquake in Luzon, the Philippines. Top: a grass hut adjacent to the Philippine fault scarp was knocked down during the earthquake but put up again without any injuries or losses of life. Bottom: Hyatt Terraces Hotel in Baguio, heavily damaged during the earthquake, with losses of life. Source: photos courtesy of T. Nakata, Hiroshima University.

the Philippine fault north of the city, south of a segment of the same fault that would rupture in 1990. Another earthquake that did damage in Manila struck on July 18, 1880. It was of magnitude 7.6 on the Philippine fault east of Manila.

An earthquake on June 3, 1863, sometimes called the Intramuros earthquake because of damage to the walled inner city of Manila, may have had its epicenter in Manila Bay because the earthquake was accompanied by a tsunami. Two earthquakes, both of magnitude 7.3, struck in 1937 and 1968. It is clear that Manila has had frequent large earthquakes in the past and will continue to experience others in the near future. The difference is that in recent years the population of Metro Manila has grown enormously, including its slums, and in the near future, an earthquake may strike that will be a catastrophe, with huge loss of life.

Not all of the source faults for some of Manila's damaging earthquakes are known. Many of Manila's earthquakes have their epicenters far away, including earthquakes on the Philippine fault. Losses from them are large because of poorly constructed buildings and a rapidly increasing population. However, there are two earthquake faults that are within Metro Manila itself: the West Marikina Valley and East Marikina Valley faults, now known simply as the Valley fault system. The north end of this fault system is close to the Philippine fault at Dingalan Bay on the east coast of Luzon. The fault bifurcates into the East Valley fault, extending into Laguna de Bay, and the West Valley fault, which extends through downtown Metro Manila, within Quezon City, and continues southward in the direction of Lake Taal and its volcano, although there is no clear evidence that it has continued as far as Lake Taal. The faults form the two sides of Marikina Valley. The present sense of displacement is by strike-slip, in which the Manila Bay side tends to move south with respect to the hills east of Metro Manila.

To learn about the paleoseismic history of the Marikina fault system, a trench was excavated by the US Geological Survey (USGS) north of Manila on a fault connecting the West Valley and East Valley faults. By dating sediments cut by the fault, the USGS geologists concluded that the fault had sustained at least four surface-rupturing earthquakes since AD 600. One of these earthquakes might have been a historical earthquake that produced damage. The most likely candidate was an earthquake on August 20, 1658; another earthquake, possibly on the Valley fault system, took place on February 1, 1771. Additional paleoseismic trenching studies now underway by PHIVOLCS has as an objective the location of source faults for these earthquakes.

Compared with large cities in the United States or Japan, the faults and earthquake hazards to Metro Manila are not well known. PHIVOLCS is engaged in an extensive trenching program to establish more clearly the earthquake hazards from individual faults, thereby leading to a better hazard assessment for all of Metro Manila. In addition, Metro Manila, assisted by the Government of Japan, has produced an analysis of the effect of earthquakes on the Philippine fault and the Valley faults on the city with an objective of strengthening critical facilities against earthquakes. However, Transparency International gives the Philippines a ranking of 34, not as low as some countries, but low enough that corruption is still a major problem in retrofitting the city.

21 Lima, Peru: Inca earthquake-resistant construction and a bogus American earthquake prediction

HISTORY

We had been at sea for two months, taking ocean-bottom cores in the Pacific Ocean off South America to learn more about the geology of the Peru–Chile trench and subduction zone, the world's longest convergent plate boundary. Now we were headed for Callao, the main port of Lima, Peru. Most of our core samples had been collected from the Nazca plate, a little-known oceanic slab that is being tectonically driven eastward beneath the continent of South America. There is a small town in southern Peru named Nazca, or Nasca, but the name of the tectonic plate comes from a mysterious set of lines drawn as much as 2000 years ago in the soil of the world's driest desert. Some of the Nasca lines are arrow-straight; others depict hummingbirds, spiders, monkeys, sharks, llamas, and lizards. They were not recognized at all until people viewed them from the air in the 1920s, and because they were not accompanied by a written record, speculation arose about their origin. Because they can be viewed only from the air, one theory was advanced that they are extraterrestrial, marking the landing sites of space ships!

Peru is a country I had always wanted to visit. The Nasca lines were one of the attractions, but another was the fact that Peru was the center of the largest pre-Columbian empire in the Americas, founded in AD 1438. The Inca empire, the largest in the world in the fifteenth century, was centered on the Andes and the plateau called the Altiplano, and it extended nearly 2500 miles (4000 km) from central Chile and Argentina north to Ecuador and southern Colombia. The Incas called this empire *Tawantin Suyu*, Land of the Four Quarters, and its capital was Cusco, at an altitude higher than 10,000 feet (3400 m) above sea level in the Andes (Figure 21.1). Much of the empire was governed by local tribes that had pledged loyalty to the Inka, as the emperor was called. The empire was more like the British Common-wealth than the Japanese empire of World War II. If the Spanish had not intervened, Tawantin Suyu might have evolved into a multicultural nation like India is today.

The Incas had no written language, relying on a complex set of knotted and colored strings called *quipu* or *khipu*, which appear to be a three-dimensional accounting system similar to decimals, together with writing that is still incompletely undeciphered, thanks to efforts by the Spanish *conquistadores* to exterminate the Inca culture completely. The Incas built a 14,000 mile (22,500 km) network of roads, many in the Altiplano, the highest plateau on Earth outside of Tibet. It is still possible to travel in the Peruvian high country on Inca roads and trails, although many of them are steep, suitable for llamas but not

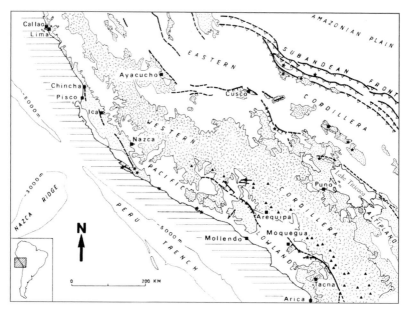

FIGURE 21.1 Map of southern Peru locating Lima, Cusco, and the Peru trench, location of the subduction zone. Crustal faults are also shown. Source: Yeats (2012, Fig. 5.9) and Sébrier *et al.* (1985).

horses or cars. One of these Inca trails takes you to the ruins of the abandoned city of Machu Picchu, lost until Hiram Bingham discovered it in 1911.

The Incas were preceded by other cultures, including the Chavín, which flourished more than 2000 years ago. This was followed by two independent countries, Wari, along the Peruvian coast and adjacent Andes, and Tiwanaku, centered on Lake Titicaca in Peru and Bolivia. The capital of Tiwanaku, also named Tiwanaku, was constructed at 12,600 feet (3840 m) in modern Bolivia, the highest major city in the world. The Inca empire was established through assimilation or conquest of older states, and it was at its height at the time of the arrival of the Spanish *conquistadores* under Francisco Pizarro in 1532.

The eleventh Inca emperor had died in Ecuador in 1525, prior to Pizarro's arrival, leading to a disagreement between two of his sons over who would be the twelfth emperor. This led to several years of civil war between Atawallpa (Atahualpa), who was then living in Ecuador, and his brother, Washkar (Huáscar), who was in the imperial capital, Cusco. Washkar was defeated, and Atawallpa was on his way to Cusco when he encountered a small force of 168 hairy, fair-skinned men riding giant animals (horses, which the Incas had never seen) in the town of Cajamarca.

Atawallpa met Pizarro's expedition in the Cajamarca town square dressed in ceremonial garb. Pizarro's force attacked the Incas with cannon fire, capturing Atawallpa and imprisoning him. Atawallpa agreed to fill a large room with gold objects to gain Pizarro's favor, but this only led to his execution. The Incas could not deal with the Spanish on horseback, although the main reason the Incas were defeated was the spread of disease introduced by the Europeans, especially smallpox. After the conquest, Spanish missionaries forced the natives to work under horrific conditions, adopt Christianity, and abandon their own ancient religion. The Spanish Viceroyalty of Peru was established in Lima, which had been founded in 1535. By 1572, the last Inca emperor, Túpac Amaru, was captured, and formal Inca resistance was at an end.

The death rate from smallpox was catastrophic. The population during the last days of Inca rule, before the arrival of the Spanish, had been 12 million. Forty-five years later, it was only 1.1 million. The Inca civilization was effectively destroyed. However, the native languages of Quechua and Aymara survived, and they are today the most widely spoken languages of Peru.

I had not been in Peru long before I recognized that there are two Perus, the ruling people of European background, and the far more numerous Quechua- and Aymara-speaking indigenous people. This was particularly noticeable in Cusco, where the Europeans ran the businesses and hotels, and the natives did the back-breaking work. I recall one man in Cusco, no more than five feet tall, carrying a huge bag of cement down the street, with no one offering to help him. The year was 1975, and I wondered how long this unequal treatment would continue before the natives had had enough. Within a decade, my question was answered, with the rise of guerrilla insurgents, the Sendero Luminoso (Shining Path) movement and the Túpac Amaru Revolutionary Movement, named for the last Inca emperor. These insurgencies led to civil unrest, brutally put down by the government. Much of the support for the rebels had come from narcotraffickers, who were expanding the cocaine industry. The cultivation of coca leaves was and is one of the major industries of Peru and is a source of income to poor native Andean farmers.

EARTHQUAKES

The first major earthquakes were described in 1533, even before Lima was founded. A stronger earthquake struck in 1555, damaging many buildings and resulting in deaths. Other large earthquakes struck in 1568, 1578, 1582, 1584, and 1586. The 1586 earthquake leveled buildings in the Lima–Callao region; a tsunami was recorded with wave heights as high as 60 feet (20 m). Some of the South American earthquakes have been correlated with "orphan" tsunamis in Japan – tsunamis not accompanied by a local Japanese earthquake (see discussion in Chapter 8 on Cascadia).

The Incas did not record earthquakes because they did not have a system of writing, although there is evidence from their oral traditions that they were aware of earthquakes. An Inca god named Pachacámac was supposed to be able to control earthquakes. The Inca response to the hazard from earthquakes was reflected in the engineering that was used in the construction of Inca buildings. They used heavy stone blocks that were fitted precisely, even though they were not mortared (Figure 21.2). The trapezoidal shape of building walls, which sloped steeply toward the outside of structures, caused them to stand against the horizontal shaking of earthquakes, even as buildings built by the Spanish were collapsing during an earthquake. The Spanish constructed buildings, including the cathedral in Lima's Plaza de Armas, the same way they did in Spain, and these buildings did not survive, whereas the Inca buildings stood up to earthquakes very well. This was a case in which buildings constructed by the "first Peru," using unwritten principles of earthquake engineering, withstood earthquakes, whereas the Spanish buildings of the "second Peru" did not. Some of these walls in Cuzco and Machu Picchu appear as fresh today as when they were built. In later years, the Spanish learned from their mistakes and built structures on top of Inca basements of buildings that had been destroyed on purpose by the Spanish.

Similarly, pre-Columbian terraced fields showed evidence of advanced civil engineering prowess. North of Cusco, I saw terraces that were sloped gently toward the hill to prevent streams from eroding the fields. These terraced fields, which were used for irrigation, are everywhere, and it left me believing that the engineering that went into their design was far superior to the engineering of their Spanish conquerors. These fields were landscaped not by the Incas but centuries before by their predecessors, the Wari.

The boundary between the South American and Nazca plates lies at the Peru–Chile trench, 100 miles off the coast of Lima (Figure 21.1). It is the longest oceanic trench on Earth, extending more than 3600 miles (5800 km) from Colombia to southern Chile. The rate of subduction is also high. Strain accumulates at a rate of 2–3 inches

FIGURE 21.2 Pre-Columbian Inca architecture at Sacsaywamán, near Cusco, Peru. Stones are fitted together without use of mortar. These outward-sloping walls are resistant to earthquakes and have stood up well, whereas walls built by the Spanish collapsed. Source: photo © Leander Canaris/Flickr, made available under an Attribution 2.0 Generic (CC BY 2.0) license at www.flickr.com/photos/28299495@N04/2663912402.

(6–8 cm) per year, fastest in the south. The trench is as deep as five miles (8 km), and mountains on the edge of the continent include Aconcagua in Chile, the highest mountain in the Western Hemisphere at just under 23,000 feet (6960 m).

The long, simple shape of the Peru–Chile trench is the reason the plate boundary has been the source of superquakes, including the largest ever recorded, in southern Chile on May 22, 1960, with a magnitude of 9.5 (see Figure 11.1). This earthquake was preceded by a somewhat smaller earthquake in 1835, experienced by Charles Darwin during the voyage of the *Beagle*. The most recent earthquake in that region of southern Chile was the Maule earthquake of magnitude 8.8 on February 27, 2010 in approximately the same area as the 1835 earthquake that had been described by Darwin (see Chapter 11 on Santiago, Chile). Other superquakes were recorded off Ecuador in 1906, with a magnitude of 8.8 (see Chapter 22 on Guayaquil and Quito), the largest earthquake of that year, much larger than the more famous San Francisco earthquake several weeks later. Two earthquakes of about the same size struck near the border between Peru and Chile in 1868 and 1877. These earthquakes produced tsunamis that killed more than 2000 people, including some as far away as Japan.

Two Peruvian earthquakes stand out as particularly devastating. An earthquake greater than magnitude 8 struck 50 miles (80 km) northwest of Lima on October 28, 1746, producing strong shaking for three to four minutes. The city at that time had a population of 60,000, of which 1141 were killed in the earthquake. Only 25 of an original 3000 houses were left standing. Afterwards, there was an attempt to limit construction to one-story buildings, but this brought opposition, and buildings of more than one story were allowed as long as they used bamboo, which is flexible, rather than adobe bricks. The 1746 earthquake was accompanied by a tsunami at least 75 feet (23 m) high that leveled the port city of Callao. Of Callao's population of 5000–6000, fewer than 200 survived.

An earthquake of magnitude 7.9 struck the Ancash region northwest of Lima on May 31, 1970. Most buildings in the city of Huaraz collapsed, killing 20,000 people. But the worst effect came from the collapse of a great slab of snow, ice, and rock from a near-vertical cliff near the summit of Peru's highest mountain, Nevado Huascarán, in that part of the Andes called the Cordillera Blanca. The mass of material fell 3000 feet (910 m), broke up and slid across a glacier, then became airborne. After returning to Earth, the mass of rock and ice became streams of debris and surface water that traveled down the valley of the Shacsha River at speeds as high as 150 miles (240 km) per hour. Some of the debris overtopped a ridge and buried the city of Yungay, and then destroyed the nearby city of Ranrahirca. The entire disaster occurred in under four minutes, from the first collapse on Nevado Huascarán to the obliteration of the city of Ranrahirca. The total loss of life from the earthquake was 80,000, one of the greatest natural disasters to strike either South or North America during historic times.

POPULATION GROWTH AND FUTURE RISK FROM EARTHQUAKES

Lima currently has a population greater than eight million. As in other parts of the developing world, much of the population has moved to the city only in recent decades, living in poorly constructed slums and shantytowns. More than 40% of Lima's residents are from rural Peru and have come to the city in search of jobs. They live on unstable soil that can liquefy in an earthquake and amplify the seismic waves, or on rickety hillside slums subject to landslides triggered by earthquakes. Their houses are commonly constructed of adobe (mud blocks), which increases the probability of collapse in an earthquake.

One difference is that because earthquakes have happened so frequently in Lima's history, people there are more aware of earthquakes than they are in many other parts of the world. The National Civil Defense Institute of Peru estimates that a magnitude 8 earthquake near Lima will kill at least 50,000 people, injure more than 680,000, and destroy 200,000 homes. The director of the Center for Disaster Study and Prevention (PREDES), a non-governmental organization headed by an architect, Juan Sato, and funded by the charity OXFAM, says that Lima is the city in South America most at risk from earthquakes. It has one-third of Peru's population, 70% of its industry, 85% of its financial services, its entire central government, and most of the country's international commerce. However, Transparency International rates the Corruption Perception Index of Peru in 2012 as 38, indicating that despite an increased awareness of deaths from poorly constructed buildings, corruption will most likely lead to a high death toll in the next earthquake.

In 2010, GeoHazards International (GHI) undertook the upgrading of a primary school building in a rural village called Chocos, in partnership with Stanford University, the Pontifical Catholic

University of Peru, and a Peruvian nonprofit organization called Estrategia. Students, teachers, and local builders are being taught how to build earthquake-resistant structures, including those with adobe. Funding is provided through the Swiss Reinsurance Company, Thornton Tomasetti Foundation, and GHI's Oyo Memorial Fund.

THE LIMA EARTHQUAKE PREDICTION

Because the inhabitants of Lima are naturally paranoid about earthquakes, the publication of an earthquake prediction for Lima by an American government geophysicist came close to causing an international incident (Olson *et al.*, 1989).

Most earthquake predictions originate from non-scientific sources, but because such predictions become an instant news story, they take on a life of their own (see Chapter 4). The Lima prediction was made by Brian Brady, a young geophysicist with the US Bureau of Mines, who specialized in mine safety. Between 1974 and 1976, Brady published a series of papers in an internationally respected scientific journal in which he argued that characteristics of rock failure leading to the collapse of underground mine walls are applicable to earthquakes. His papers combined rock physics and mathematical models to provide what he claimed was an earthquake "clock" that could predict the precise time, place, and magnitude of a forthcoming earthquake. Brady pointed out that earthquakes near Lima in 1974 and 1975, including the October 3, 1974 Lima–Callao earthquake of magnitude 7.2 in which 179 people died, had occurred in places where no earthquakes had occurred in a long time, and he predicted a much larger earthquake off the coast of Peru. His prediction was supported by one of his colleagues, William Spence, a respected geophysicist with the USGS. However, the prediction did not lead to acceptance by other American earthquake scientists.

His prediction received little attention at first, but gradually it became public, first in Peru, where the impact on the rapidly growing population of Lima would be enormous, then in the United States, where various federal agencies grappled with the responsibility of endorsing or denying a prediction by a government scientist that had very little support among mainstream earthquake experts.

The prediction became a major media story when Brady announced that the earthquake would strike Lima on June 28, 1981, and would have a magnitude of 9! The Peruvian government asked the US government to evaluate the prediction that had been made by one of its own scientists. In response, a meeting in January 1981 of a group of senior earthquake scientists called the National Earthquake Prediction Evaluation Council (NEPEC) reviewed the prediction so they could make a recommendation to the director of the USGS on how to advise the President of the United States and the Peruvian government. NEPEC evaluated the Brady and Spence prediction – and rejected it.

Did the rejection by NEPEC end the controversy? Not at all, and Brady himself was unconvinced that his prediction had no scientific merit. Charles Osgood of *CBS News* interviewed Brady shortly after the January meeting of the council, but the interview was not made public until June, shortly before the predicted earthquake was supposed to take place, thereby enhancing its news value. Officials of the Office of Foreign Disaster Assistance in the State Department took up Brady's cause. The press labeled the NEPEC evaluation as a "trial and execution" and described the NEPEC panel as a partisan group that would rather destroy the career of a dedicated scientist than take a risk and endorse his prediction.

John Filson, a senior official with the USGS Earthquake Hazards Reduction Program, made a high-profile public visit to Lima so he would be there on the day of the predicted earthquake to reassure the public. No superquake arrived to keep Brady's appointment with Lima, and none has arrived there to this day.

But there were damaging earthquakes after the Brady prediction, occurring with a frequency similar to earthquakes earlier in Peru's recorded history since the early sixteenth century. On June 23, 2001, an earthquake of magnitude 8.4 struck southern Peru close to an earlier earthquake of magnitude 7.7 that took place in 1996 and also close to the epicenter of a superquake in 1868. The cities of Arequipa and Tacna (Figure 21.1) were strongly affected. At least 75 people died, including 26 killed and 64 missing in the accompanying tsunami. On August 15, 2007, the Ica–Pisco earthquake of magnitude 8 had its epicenter close to the city of Pisco, 90 miles southeast of Lima (Figure 21.1). There were 519 lives lost, with 430 killed in Pisco alone, near the epicenter. Despite Peru's extensive experience with earthquakes, the response was chaotic, with a lack of coordination among the military, the private sector, international non-governmental organizations, and the United Nations. An earthquake of M_w 8.2 struck near Iquique in northern Chile on April 1, 2014, preceded by numerous foreshocks.

None of these earthquakes struck close to Lima, as required by the Brady prediction, and both were much smaller than predicted, so they did not meet the criteria of a successful prediction. Remember that the magnitude scale is logarithmic, and a magnitude 9 would be many times larger than a magnitude 8.

Will there be a superquake of magnitude 9 at Lima? Probably, because the Peru–Chile subduction zone has produced several of that size in its recorded history. But Brady's prediction added no information that would lead the Peruvian government to raise the alarm. Like other future earthquakes discussed in this book, the earthquake could happen tomorrow or a century from now.

The reason for the problem is our scientific failure to be able to tell the people of Peru that an earthquake of a given size would strike at a given place and a given time. That is still the main question Peruvians want answered, and the scientific community was unable to rise to the challenge. That was true in 1981 when Brady announced his prediction, and it is still true today.

22 Andean earthquakes in Quito and Guayaquil, Ecuador

INTRODUCTION

Ecuador is the smallest Andean country in South America, bounded by Colombia on the north and Peru on the south. The country was founded as the Kingdom of Quito in AD 980, which was later conquered by an Inca army from Peru in 1462. The Inca occupation lasted until 1534, when the Inca empire was defeated by the Spanish. The Spanish occupation was similar to that in other parts of South America: brutal enslavement of Ecuador's inhabitants and forced conversion to Catholicism. Ecuador became part of the Viceroyalty of Peru in 1563. The country remained under the rule of Spain until it was liberated by Simón Bolívar in 1822.

The two largest cities, Quito and Guayaquil, were founded in the sixteenth century. The capital city of Quito is located in a high Andean valley (Depresión Interandina, or Inter-Andean Depression; Figure 22.1) at an elevation of 9350 feet (2800 m), whereas Guayaquil, the main port of Ecuador, is on the banks of the Guayas River near the shores of the Gulf of Guayaquil. Quito's population was 10,000 in the last days of Spanish rule, and it has grown to more than two million today. Guayaquil had 2000 inhabitants in AD 1600, but at the present time its population is more than 2.3 million, making it the largest city in Ecuador, larger than Quito, the national capital.

GUAYAQUIL AND COASTAL AND OFFSHORE ECUADOR

Both Guayaquil and Quito are subject to earthquakes, but the earthquakes endangering Guayaquil are far different from those affecting Quito. Guayaquil's hazard is from the subduction zone driving the oceanic Nazca Plate east-northeastward beneath the South American continent at a rate of nearly 3 inches (7 cm) per year. A complication is that the Nazca oceanic plate is not a smooth slab but contains a mountainous highland called the Carnegie Ridge that is colliding with the continent (Figure 22.1). The highest mountains on the Carnegie Ridge rise up from the ocean floor above sea level as the Galápagos Islands, themselves a dependency of Ecuador. The Carnegie Ridge intersects the South American continent west of low coastal hills northwest of Guayaquil. All of the large subduction-zone earthquakes have affected coastal Ecuador and Colombia north of the city and north of the Gulf of Guayaquil.

The year 1906 is famous for being the year of the great earthquake that destroyed San Francisco, California; but two and a half months earlier, an earthquake many times more powerful struck the subduction zone off the coast of Ecuador and Colombia. This was the Esmeraldas earthquake of January 31, 1906, a superquake of magnitude 8.8. The town of Esmeraldas is on the northern Ecuadorian coast, near the Colombian border, far from Guayaquil. This earthquake resulted in 1000 deaths, a number that was not larger because of the small population at that time along the coast of Ecuador and Colombia.

The earthquake generated a tsunami that struck the coast 30 minutes after the mainshock, causing 500–1500 fatalities. The tsunami crossed the Pacific Ocean 19.5 hours later and was recorded in Japan as an "orphan tsunami," meaning a tsunami not accompanied by an earthquake. About 400 miles (640 km) of the subduction zone underwent rupture, but the southernmost edge of the rupture was far enough north of Guayaquil that it did not cause many fatalities in that city. If the rupture zone had extended 100 miles (160 km) farther south, the loss of life in Guayaquil would have been much higher.

The Esmeraldas superquake was followed 36 years later by an earthquake of magnitude 7.8 on May 14, 1942 that affected only the southern part of the 1906 rupture zone in Ecuador. This earthquake killed 300 people, about half of them in Guayaquil. Concrete block

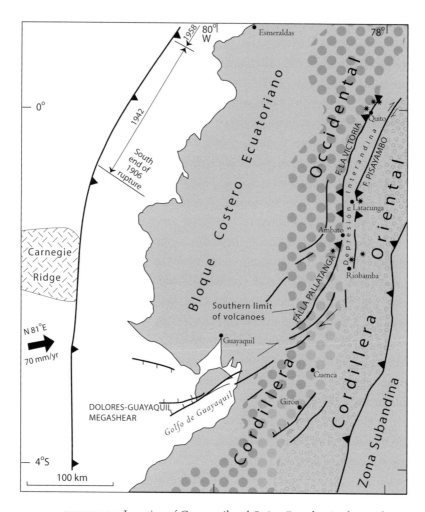

FIGURE 22.1 Location of Guayaquil and Quito, Ecuador, in the northern Andes of South America. Guayaquil is east of the subduction zone between the Nazca and South American plates; the southern edge of the 1906 Esmeraldas superquake of magnitude 8.8 is shown. Quito is in a mountain valley bounded by reverse faults, several of which have generated damaging earthquakes since the arrival of the Spanish. Asterisk symbol: active volcanoes. Source: modified from Yeats (2012, Fig. 5.6) and Winkler *et al.* (2005).

buildings collapsed, including the US consulate; the American consul was killed in the earthquake. Another earthquake of magnitude 7.8 struck the next segment to the north on January 19, 1958, and a third earthquake of magnitude 7.7 struck the next zone to the north, off the coast of Colombia, on December 12, 1979. The rupture zones of the three earthquakes taken together were about the same length as the single rupture zone of the 1906 superquake. It was thought at the time that even though the three earthquakes filled the 1906 rupture zone, the energy released by the three later earthquakes was much smaller than that from the single 1906 superquake, but this interpretation is now controversial.

Could there be a large earthquake on the subduction zone south of the 1906 rupture? The southern end of the 1906 and 1942 rupture zone was close to the northern edge of the Carnegie Ridge, colliding with the continent northwest of Guayaquil. The Carnegie Ridge and the subduction zone immediately to the south have not been subjected to a major subduction zone earthquake during Ecuador's recorded history. Does this mean that Guayaquil does not face a threat from a subduction zone earthquake similar to the Esmeraldas earthquake, or is this part of the subduction zone building up for a Big One?

A smaller earthquake of magnitude 5.9 on August 18, 1980 struck the subduction zone at 35 miles (56 km) depth, south of the extension of the Carnegie Ridge and relatively close to Guayaquil, causing the deaths of 11 people in addition to injuries. An earthquake northeast of Guayaquil of magnitude 6.8 about 80 miles (130 km) beneath the surface on November 16, 2007 did not cause deaths or injuries, in part because it was relatively deep. These earthquakes were on the eastern extension of the Carnegie Ridge.

Aside from the 2007 earthquake, the subduction zone south of the Carnegie Ridge appears less active than the zone to the north, but this observation alone does not mean that the subduction zone off southern Ecuador could not generate a large earthquake. Other subduction zones, most notably Cascadia, are seismically quiet, but paleoseismic work has identified Cascadia as a potential earthquake source (see Chapter 8). The investigative work has not been done in Ecuador that would identify the southern coast of Ecuador as a major earthquake hazard – or not.

The Esmeraldas superquake of 1906 and the three earthquakes that followed it indicate that the Nazca plate is continuing to drive beneath the edge of South America. Aside from the Carnegie Ridge, there is no tectonic boundary within the Nazca plate between the northern and southern parts of the subduction zone, implying that if there is shortening off the coast of northern Ecuador, there should also be shortening to the south. In support of this idea, an earthquake in 1953 of magnitude 7.5 occurred beneath the southern shore of the Gulf of Guayaquil. Could this shortening be expressed by earthquakes in the Andean crust, including the region around Quito?

THE ANDES: EARTHQUAKE HAZARD TO QUITO

The Andes in northern Ecuador consists of two parallel ranges, the Cordillera Occidental and Cordillera Oriental (Western and Eastern Mountain Ranges), separated by the Inter-Andean Depression, which consists of several basins in which the major cities are located, including the capital city of Quito. These ranges are bounded by faults with a sense of right-lateral strike-slip and reverse-slip, reflecting the oblique convergence direction between the oceanic Nazca and continental South American tectonic plates. The slip rate on range-bounding faults is too slow to explain the small number of historical subduction zone earthquakes south of the rupture zone of the 1906 Esmeraldas superquake. One of the best known of the Andean faults, the Palla-tanga fault (Figure 22.1), has a slip rate of about 0.2 inches (5 mm) per year, a small fraction of the convergence rate between the Nazca and South American plates. These crustal faults are the sources of earthquakes that have caused major damage to Andean cities in Ecuador.

The Andes of northern Ecuador also has major active volcanoes, including Cotopaxi at 19,347 feet (5897 m). Quito itself is at the base of an active volcano, Pichincha, which produced a swarm of earthquakes in 1998–1999. Quito is crossed by an active reverse fault. The pre-Columbian earthquake history of Quito was analyzed by a French paleoseismologist, Christian Hibsch, who found evidence of 28 earthquakes in the 1500-year period prior to historical record-keeping. An earthquake between the tenth and sixteenth centuries AD had intensities high enough to have ruptured the entire Quito fault system. Historical earthquakes in 1797 and 1868 producing major damage in the Inter-Andean Depression are described further below.

An earthquake on February 4, 1797 that destroyed the city of Riobamba, south of Quito, was visited by the German explorer Alexander von Humboldt in 1801–1802. Using earthquake intensities, the magnitude was estimated as 7.6, although the damage suggests the rupture was no more than 40 miles (64 km) long. The losses of life were estimated as at least 25,000 people, although higher estimates were given by von Humboldt. The epicenter was close to the Pallatanga right-lateral strike-slip fault, although no surface rupture has been identified on that fault from the 1797 earthquake. Riobamba is close to the active Tungurahua volcano, 16,486 feet (5025 m) high, in the Cordillera Oriental. This volcano entered an active eruptive phase in 1999, continuing to the present. Farther south, the Ecuadorian Andes and northern Peruvian Andes lack active volcanoes.

North of Quito, large earthquakes struck on August 15 (El Ángel, magnitude 6.6) and August 16, 1868 (Ibarra, magnitude 7.25), in which up to 25,000 people lost their lives. These earthquakes may have struck faults trending north-northeast with a right-lateral and reverse sense of displacement. This means that during the earthquake, the country west of the faults moved to the north relative to country to the east, in addition to the west and east sides moving toward each other. The El Ángel earthquake on August 15 was close to the Colombian border. The Ibarra earthquake on August 16 was closer to Quito, and people in the Quito area were killed during this earthquake. The city of Ibarra was destroyed by the earthquake and was abandoned for four years. People returned to Ibarra in 1872, and the date of the earthquake is remembered every year by residents of this region.

The Quito fault system itself generated a magnitude 6.4 earthquake in 1587, but at that time, the main part of the city was about 15 miles (25 km) south of the epicentral area. Today, the city has expanded beyond the 1587 epicenter. Smaller events have jolted the city throughout its history. The last one that killed people was in 1990 (four casualties) in the same area as the 1587 earthquake. GPS measurements show that the Quito fault may be moving at about one-sixth inch (4 mm) per year. GPS evidence shows that the fault has accumulated strain for the past 18 years and is locked at about 2–4 miles (3–6 km) depth beneath the city.

Active volcanoes are part of the Andes between southern Colombia and Riobamba in central Ecuador. This means that the residents must be concerned about both earthquake and volcanic hazards. Quito, the largest city in the region, has not experienced earthquakes during its short history as damaging as the earthquakes to the north in 1868 and to the south in 1797, although the paleoseismic evidence shows that earthquakes of this size are to be expected there. Because more than two million people live in Quito, losses from the next earthquake are likely to be very heavy.

South of Riobamba, the Pallatanga fault curves toward the southwest and enters the Gulf of Guayaquil as a west-trending fault called the Dolores–Guayaquil Megashear (Figure 22.1). Whereas the Pallatanga is a strike-slip fault, the faults of the Megashear are normal faults in which the north side is moving away from the south side. The Megashear intersects the subduction zone south of its intersection with the Carnegie Ridge. Although this structure is considered by scientists as one of the most important in South America, it has not been struck by a major earthquake during the past five centuries.

Will the small country of Ecuador be able to respond to a major earthquake disaster? Fortunately, the lowlands of Ecuador east of the Andes contain large deposits of oil and gas that result in membership of Ecuador in OPEC (Organization of Petroleum Exporting Countries). This means that, like Venezuela, Ecuador has the financial means to respond to a major catastrophe. However, the relation between the government and the petroleum industry has not been smooth because of charges of pollution of the environment east of the Andes by oil production. In addition, oil refineries, storage tanks, and export facilities are located close to the epicenter of the 1906 Esmeraldas superquake, and the pipeline bringing the oil from the eastern lowlands to the coast crosses several major active faults mapped by Ecuadorian and foreign scientists. In 1987, an earthquake of magnitude 7.1 occurred 50 miles (80 km) east of Quito and destroyed the pipeline over a distance of more than 25 miles (40 km), stopping the export of oil for five months.

However, action is being taken about the earthquake hazard to schools in Quito, Ecuador's capital city. In December 1994, GHI (Geohazards International), the nonprofit consulting firm founded by seismologist Brian Tucker (see the section on Kathmandu in Chapter 18), joined with Kunio Suyama of Oyo Corporation and Jean-Luc Chatelain of ORSTOM to launch the Quito School Earthquake Safety Project. This project was overseen by two committees, a Policy Advisory Committee composed of local government officials responsible for school construction, and a Technical Advisory Committee made up of non-Ecuadorians familiar with school retrofits and safety. GHI also worked with my colleague, Professor Hugo Yepes of the Escuela Politécnica Nacional (National Polytechnic School) of Ecuador and with the University of British Columbia.

Fifteen schools were selected for special study, based on different types of construction and on high use and vulnerability. Funding has been obtained locally and with assistance from USAID to retrofit ten of the school buildings. The committees determined that retrofitting the schools could be done relatively inexpensively, thereby generating support for local funding. The school retrofits will be used to upgrade schools throughout Ecuador. National legislation has been adopted to strengthen building codes for public and private buildings.

Sadly, Dr. Suyama of Oyo Corporation died before the project was to begin. This chapter is dedicated to his memory.

23 Caracas: lots of oil, but little interest in earthquakes

HISTORICAL BACKGROUND

The capital of Venezuela is in some respects a modern, wealthy city, in part due to the availability of petrodollars. Venezuela has the largest oil reserves in Latin America, and its reserves are in the top ten worldwide. The country exports a large amount of its oil to the United States, despite their political differences. Oil provides one-third of Venezuela's gross national product and 50% of income to run the national government. The recent drop in the price of oil has had a major impact on the Venezuelan economy.

The city of Santiago de León de Caracas was founded by the Spanish in 1567 and became the capital of the Province of Venezuela in 1577. The city occupies a mountain valley and is separated from the coast by a mountain range, the Cordillera de la Costa, as high as 7400 feet (Figure 23.1). The reason Caracas was not located on the nearby Caribbean coast was the fear of attack by English fleets or by pirates; the Cordillera de la Costa served as a wall against invasion. Nevertheless, pirates breached the mountain wall and burned the city in 1595.

On June 11, 1641, Caracas was destroyed by the St. Barnabas Day earthquake at a time when it had only a small population. Five hundred people were killed. There was a proposal to move the capital to a safer place, but this plan was abandoned when it was not supported by the Spanish authorities.

The successes of the American and French revolutions led to a desire by Venezuelans to be free of Spanish rule. The first revolution was put down by the Spanish in 1797. The revolution that was ultimately successful began under the leadership of Venezuela's George

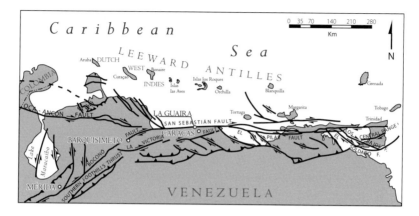

FIGURE 23.1 Fault map of Venezuela, showing that the major cities of the country are crossed by seismically active faults, including the plate-boundary Boconó and San Sebastián faults. Source: modified from Yeats (2012, Fig. 4.13) and Audemard *et al.* (2000).

Washington: Simón Bolívar, known as El Libertador, when he declared Venezuela's independence in 1811.

THE 1812 EARTHQUAKES

On March 26, 1812, the year after Venezuela declared its independence from Spain, disaster struck. Caracas was destroyed by an earthquake of magnitude 7.7 to 8, much larger than the previous earthquake of 1641. There were two shocks separated by nearly 45 minutes. The first destroyed Caracas, and the second destroyed the city of Mérida, in the Andes Mountains in the far western part of the country (Figure 23.1). Several other cities were also destroyed, including Barquisimeto, northeast of Mérida, and La Guaira, the port city for Caracas. There were 26,000 lives lost, between 5% and 10% of the population of Venezuela at that time.

The earthquakes struck during Holy Week on the Catholic holiday of Maundy Thursday, and the archbishop of Caracas, speaking for the Spanish authorities in response to the rebellion led by Bolívar, referred to it as "the terrifying but well-deserved earthquake" that "confirms in our days the prophecies revealed by God to men about the ancient impious and proud cities: Babylon, Jerusalem and the Tower of Babel." As in 1641, a suggestion was made to move the capital to a safer place, but as before, the suggestion was rejected. Despite attempts by Spanish religious authorities to blame the earthquakes on the revolution, the war continued, and after a victory in 1821 by Simón Bolívar, Venezuela finally achieved its independence.

It has not yet been possible to assign the 1812 earthquakes to a specific fault, but the distribution of devastated cities follows an active seismic zone that includes the boundary between the South American and Caribbean tectonic plates. The most active fault along this plate boundary (and a possible source of one of the 1812 earthquakes) is the northeast-trending Boconó strike-slip fault, which is the boundary between the Caribbean region, moving relatively east, and the main part of South America, moving relatively west (Figure 23.1). The Boconó fault is close to the city of Mérida, accounting for the major destruction suffered by that city in 1812. Over thousands of years, the Mérida Andes mountain range has been uplifted, accompanied by strike-slip movement on the Boconó fault (Audemard et al., 2000).

OTHER EARTHQUAKES

Northeast of Barquisimeto, the plate-boundary fault turns eastward as the San Sebastián fault, which extends north of the Cordillera de la Costa near the port of La Guaira (Audemard et al., 2000; Figure 23.1). The San Sebastián fault is the source of earthquakes affecting Caracas, possibly including the earthquake of 1812, a problem for scientific investigation because the fault is offshore and unavailable for direct view except at the Caracas International Airport at La Guaira. However, farther east, the plate-boundary fault comes ashore as the El Pilar strike-slip fault (Figure 23.1). This fault was the source of surface-rupturing earthquakes in 1530 (first recorded earthquake in Venezuela), 1684, 1853, 1929, 1974, and 1997. The plate-boundary fault continues farther east across the Gulf of Paria to the island of Trinidad.

One branch of the San Sebastián fault does extend onshore into metropolitan Caracas: the Tacagua-Ávila fault, which has evidence of horizontal and vertical displacement. This fault system shows evidence of activity, but despite study by Venezuelan scientists, there is not yet a "smoking gun" that would demonstrate that it is active and capable of generating an earthquake within the city.

The Tucacas earthquake of September 12, 2009 struck the offshore San Sebastián fault east of the place where it turns to the southwest as the Boconó fault. An earlier earthquake on April 28, 1894 heavily damaged towns over a 40 mile distance southwest of Mérida and took 350 lives. Still earlier earthquakes had struck this region in 1610 and 1674. Both the 1610 and 1894 earthquakes were shown by Venezuelan scientists to have been accompanied by surface rupture on the Boconó fault; the 1894 earthquake was called the Great Earthquake of the Venezuelan Andes. The most recent earthquake to affect Caracas occurred 65 miles (100 km) west of the city on September 13, 2009, injuring 14; its magnitude was 6.4. An earlier earthquake caused damage in the city in 1967.

The historical earthquake record is clear: earthquakes have struck all along the plate-boundary fault from the Colombian border east to the Gulf of Paria. The worst of these struck in 1812, during the War of Independence, destroying Caracas as well as other major Venezuelan cities close to the plate boundary.

The most important contrast between the historical earthquakes and the next large earthquake to strike Caracas in the future is the population in harm's way. Throughout most of its history, the population of Venezuela's cities, including Caracas, was relatively small. However, the megacity of Caracas is now the home of 8.3 million *caraqueños*, as reported in 2012. If nearby cities are included, the metropolitan area has a population of 18.2 million, almost two-thirds of the population of the entire country. Much of the growth is related to the nationalized oil industry; many people have migrated to Caracas to find jobs.

One encouraging note is that, many years ago, the Venezuelan government formed an organization called FUNVISIS, which stands for *Fundación Venezolana de Investigaciones Sismológicas*, which has the responsibility of evaluating and reducing hazards from earthquakes. This organization, partly funded by the petroleum industry, contains many highly dedicated scientists who are world leaders in earthquake studies. One of them, Dr. Franck Audemard, has taken a leadership role in earthquake hazards throughout South America, filling the role formerly played by the late Dr. Carlos Schubert, whose contributions to the understanding of the Boconó fault have been used in the investigation of other strike-slip faults around the world (Audemard *et al.*, 2000). I have learned much from both of these geologists, and their work has been the basis for my analysis here.

EARTHQUAKE HAZARD REDUCTION VS. POLITICS: THE LEGACY OF HUGO CHÁVEZ

My concern is not with the skill and dedication of local Venezuelan scientists. It is, instead, whether their recommendations for strengthening Caracas and other cities against the inevitable next earthquake will be acted on in a timely manner by the federal government. Thanks to oil income, Venezuela has the money to strengthen its cities against earthquake destruction, and FUNVISIS is staffed by scientists fully qualified to advise the government about earthquake hazards. However, I saw no mention of earthquakes in political commentary coming out of Venezuela during the late President Chávez's reelection campaign. (On the other hand, earthquakes played no role in the 2012 US presidential election campaign either, so this lack of interest on the part of government leaders, as well as the general public, is not unique to Venezuela. Read the discussion in Chapter 8 on the Cascadia subduction zone in the northwestern United States.) At the present time, seismic retrofits are even less likely because of the great reduction in the price of oil, which affects the entire Venezuelan economy.

Although there is great wealth in Caracas, the city also has hillside shantytowns called *ranchos*, which suffered much of the damage in an earthquake in 1967. Poor construction and crowded conditions are a recipe for a catastrophic repeat of the 1812 earthquake, differing in the astronomical increase in population between 1812 and today. The losses in an earthquake as large as the 1812 earthquake are likely to be in the hundreds of thousands. Is the present socialist government ready for such a disaster? This question is considered next, based in part on a recent report by Jon Lee Anderson in the January 28, 2013 *New Yorker.*

For several decades after a military dictator, Marcos Pérez Jiménez, was overthrown, Venezuela was a dynamic, stable democracy. As an oil-rich nation, Venezuela had a growing middle class and a high standard of living, except for the hillside shantytowns. Caracas was an attractive, modern city with a great university, the Universidad Central de Venezuela. The city was a beautiful place for beautiful people, with glorious weather. Caracas opened the first line of its new subway in 1983. I was in Caracas in 1992 for a field trip on the Boconó fault with my friend Carlos Schubert, and I found it a delightful place to live and work. Caracas became one of my favorite cities.

David Brillembourg, a banker who made a fortune during the oil boom, began building the Confinanzas skyscraper complex, which included a 45-story skyscraper with mirrored glass. After Brillembourg's death in 1993 and the collapse of many of Venezuela's financial institutions, this skyscraper became known as the Tower of David (Figure 23.2).

The Caracas I remember no longer exists. Modern Caracas has one of the world's highest homicide rates, and drug addiction is high. Wealthier neighborhoods are fortresses, surrounded by high walls, powered wiring, and armed guards. In the early days of the Chávez regime, the abandoned, partly completed Tower of David was taken over by squatters (*invasores*) ruled by a former convict and later a born-again evangelist preacher and rabid Chávez supporter. The Tower of David became the world's tallest slum, visible throughout

FIGURE 23.2 The Tower of David in Caracas, Venezuela, built by investor David Brillembourg, who lost his money before the 45-story skyscraper could be completed. The tower was taken over by squatters (*invasores*) so that it is the world's first vertical slum. It is likely to undergo severe damage and loss of life when Caracas experiences its next major earthquake. Other buildings in Caracas are being taken over in the same way. Source: photo by Azalia D. Licón Sandoval.

the city (Figure 23.2). The Chávez government went along with this takeover as a peculiar example of its form of socialism: Find an abandoned building you like, and move in. Hundreds of buildings in Caracas were taken over in this way. Most are ruled by violent criminal elements, and it would be dangerous to enter one of these buildings.

The hillside slums were not upgraded as Chávez had earlier promised, and the slums expanded to other buildings in the city.

The problem is that the unreinforced brick construction of the shanty-towns has been extended to add-ons to the buildings taken over by *invasores*. This has simply expanded the number of *caraqueños* living in buildings subject to collapse in even a moderate-size earthquake. If these buildings, in addition to the shantytowns, are ruled by criminals, what chance would these structures have of being reinforced by engineers against earthquakes? They are too dangerous to enter.

In response to these developments, Transparency International gives Venezuela a corruption perception ranking of 19, the same value as Haiti, the poorest country in the Western Hemisphere.

It is unfortunate that a country with highly skilled earthquake scientists and engineers has been unable to fortify Venezuela's capital city against the next great earthquake. FUNVISIS geologists have investigated the Tacagua-Ávila fault, within the city limits of Caracas, and it has described its seismogenic potential. Protests following the death of President Chávez suggest that it is not clear that the government has responded to the earthquake hazard identified by FUNVISIS to strengthen this area of the fault against earthquakes. Probably not.

A repeat of the 1812 earthquake could and probably will take the lives of hundreds of thousands of residents of Caracas, with the greatest losses to the poor people living in shantytowns and the buildings taken over by *invasores*. What will it take to draw the attention of the post-Chávez government to this problem to prevent the destruction of its capital? So far, the new government has shown no more interest in protecting its citizens against earthquakes than the Chávez government did.

24 Haiti, which lost its gamble, and Jamaica and Cuba (not yet)

INTRODUCTION

The Greater Antilles region of the Caribbean is composed of island countries of moderate size that are close to, or are on, tectonic plate boundaries (Figure 24.1). These countries contain large, rapidly growing cities that are considered to be earthquake time bombs. One of these time bombs went off in January 2010. Port-au-Prince, the capital city of Haiti, was destroyed in the Léogâne earthquake of magnitude 7. Although help from the international community was sent to Haiti after that earthquake, it has not been enough. Much of the city is still in ruins, with many people still living in tents and damaged houses five years after the earthquake. But Port-au-Prince is not alone in having an earthquake problem. Other nations, including Jamaica and the Dominican Republic, have heavily populated cities close to active faults but do not have the resources to repair these cities if they are destroyed in an earthquake. Cuba has a better-developed social services system, but the southeastern end of the island also includes earthquake time bombs. But the fault that is the source of future Cuban earthquakes is offshore, and the earthquake problem is not a priority for the Cuban government.

HAITI

Haiti was not always the basket case of the Caribbean. In the late eighteenth century, the French colony of Haiti, also called Saint-Domingue, provided 40% of the sugar and 60% of the coffee imported by Europe. It was so rich that it had become known as the Pearl of the Antilles, with the industry supported by thousands of African slaves ruled by a small aristocracy of white French plantation owners. Haiti,

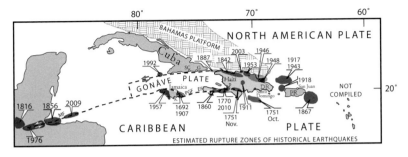

FIGURE 24.1 Location of earthquakes affecting the Greater Antilles, including Cuba, Jamaica, Haiti, Dominican Republic (DR), and Puerto Rico (PR). These earthquakes occur along one of two tectonic plate boundaries between the North American plate to the north and the Caribbean plate to the south. The intervening plate is the Gonâve plate. Other abbreviations: EF, Enriquillo fault; G, Guantánamo; K, Kingston, Jamaica; MF, Motagua fault in Guatemala, struck by an earthquake in 1976; P, Port-au-Prince; PGF, Plantain Garden fault; SC, Santiago de Cuba. Santiago and Guantánamo, including the American prison, are close to the active Oriente fault, immediately offshore. Source: modified from Yeats (2012, Fig. 4.2) and Geological Society of America Special Paper 326.

after its revolution, even provided assistance to Simón Bolívar in his drive in the early nineteenth century for independence of Venezuela and other South American countries from Spain.

Following a successful slave revolution at about the same time as American independence and the French revolution, the country became mired in conflict and mismanagement, so that by 2010 there were very few social services, and little money to deal with disasters such as hurricanes and epidemics. But Port-au-Prince had another, hidden problem. The active Enriquillo fault extends east–west, just south of the city, and is a known generator of earthquakes. The Enriquillo fault was known to be a time bomb that could rupture in a devastating earthquake at any time.

At the beginning of 2010, I had been interviewed by Katie Harmon of the magazine *Scientific American*, and I had pointed out the Port-au-Prince problem: overcrowded capital of a desperately poor country located close to a major active fault, which made it a time bomb. One week after my interview with Harmon, on January 12, the time bomb went off. (No, I didn't predict it.) The Enriquillo fault system was the source of the magnitude 7 Léogâne earthquake, with its epicenter 16 miles (25 km) west of the capital and its focal depth six miles (10 km) beneath the surface (Figure 24.1). As pointed out in the Preface, my interview went viral. As many as 230,000 people were killed outright (the exact number is controversial), another 300,000 injured (book cover), a million were left homeless, and a tsunami was generated. Also destroyed were the presidential palace (Figure 24.2), the parliament building, and the national cathedral. In a country plagued by unstable government, corruption, and poverty, Nature had played the trump card.

I pointed out to Katie Harmon that earthquake scientists were unable to predict the date of the Léogâne earthquake. However, we were able to say *when* the last big earthquake had struck close to

FIGURE 24.2 Haiti presidential palace four days after the 2010 earthquake that heavily damaged it. Source: photo by Roger Bilham, University of Colorado.

Port-au-Prince. Actually, there were two earthquakes damaging the city, one in 1751 and a larger earthquake in 1770 (Figure 24.1). This seems a long time ago to be using this information to address present-day earthquake hazards. Haiti, like the American colonies, had not yet gained its independence. Port-au-Prince only became a city in 1749, two years before the first destructive earthquake, and it became the capital of Haiti in 1770, just before the second earthquake.

Haiti became totally independent only in 1804, and from that time until 2010, the capital city had not suffered an earthquake large enough to do severe damage to the economy. The interval after the eighteenth-century earthquakes was long enough that earthquakes were not a concern to the population or to the government in power. But the two and a half centuries between the eighteenth-century earthquakes and 2010 are no more than an instant in geologic time in calculating the return period of earthquakes on the same section of a fault. Crustal strain had been accumulating that entire time on the Enriquillo fault.

The November 21, 1751 earthquake, with its magnitude currently estimated as 6.6, was not as large as the 2010 earthquake. All but one masonry building in Port-au-Prince collapsed. The earthquake had been preceded by an offshore earthquake of magnitude 8 in the preceding month, on October 18 (Figure 24.1), but that earthquake south of the Dominican Republic was far enough away that it did not do major damage in Haiti. Port-au-Prince was struck by a third earthquake, probably of magnitude 7.5, on June 3, 1770, 19 years after the two 1751 earthquakes. This earthquake was accompanied by a tsunami much larger than the one generated by the 2010 earthquake. Both the 2010 and 1770 tsunamis may have been generated by submarine landslides. Two hundred people lost their lives in the 1770 earthquake and tsunami, but the earthquake was followed by a famine that took an additional 30,000 lives. It was an inauspicious beginning for the new colonial capital of Haiti.

But these earthquakes gave us important information: the repeat time of major earthquakes endangering Port-au-Prince, about two and a half centuries. Current planners for the next earthquake might draw some comfort from the long recurrence interval, except for the fact that the earthquake in 1770 was several times larger than the earthquake of 2010, even though the 2010 earthquake caused much greater losses. The November 1751, November 1770, and January 2010 earthquakes were all related to the Enriquillo fault system, which bounds the Gonâve tectonic plate to the north and the Caribbean plate to the south (Figure 24.1). This fault had been mapped as a strike-slip fault prior to the 2010 earthquake, and it was known that the Gonâve tectonic plate north of the fault, including Port-au-Prince itself, is moving west with respect to the Caribbean plate to the south.

But, oddly enough, despite a detailed search by geologists, there was no evidence of surface fault rupture after the 2010 earthquake. The seismic signature of the earthquake, especially its aftershocks, indicated previously unknown faults that did not reach the surface at all. These faults had a different kind of motion, with one side moving up with respect to the other. The Gonâve tectonic plate moved toward the Caribbean plate, squeezing the crust between the two plates near the Enriquillo fault. In other words, the earthquake that actually destroyed Port-au-Prince in 2010 was not the earthquake that experts had warned about and was not a clone of the previous earthquake that had struck the area in 1770.

This raises the concern that the 2010 earthquake, catastrophic though it was, did not relieve the strike-slip strain that had built up along the plate boundary at the Enriquillo fault. The 2010 earthquake was much smaller than the 1770 earthquake, and it did not produce surface rupture on the main plate-boundary Enriquillo fault. The worst may be yet to come.

Another observation leading to this doomsday interpretation is the pattern of the eighteenth-century earthquakes (Figure 24.1). The first and largest earthquake was offshore from the Dominican Republic. The next two were to the west, near Port-au-Prince, but the first one was east of the city and the second one was west of it, meaning that even though the earthquakes both destroyed the city, they were on different parts of the plate-boundary fault. A fifth, smaller earthquake struck still farther west on the Tiburon Peninsula in 1860, but still on or close to the Enriquillo fault.

This leads to the possibility that the 2010 earthquake is not the end of the modern earthquake story, since the earlier earthquakes were part of a sequence of earthquakes migrating westward, like buttons ripping off a shirt. If the next earthquake occurs immediately west of the 2010 earthquake, it would still be close enough to Port-au-Prince to do more major damage to the metropolitan area. There is no evidence that the 2010 earthquake will not be followed by others in the near future, especially in view of the fact that the 2010 earthquake was not as large as the second of the eighteenth-century earthquakes on the Enriquillo fault that preceded it.

So the 2010 earthquake was not the largest one to strike Haiti in its 500-year history, but it was the most damaging. Why? The reason is population. Accurate population figures are hard to come by, but one source gives the population of Port-au-Prince in 2003 as 704,000, but in 2010 it had grown to 2.1 million. Other figures are as high as 2.5–3 million. The large loss of life and number of injuries can be related to the huge growth in population, with many people living in substandard housing in the slums of Port-au-Prince.

Some of my colleagues who have done earthquake research in Haiti for many years were worried enough about the Enriquillo fault prior to the 2010 earthquake to seek an audience with government officials, including the prime minister, to warn them about a possible earthquake threat to Port-au-Prince. They described the 1751 and 1770 earthquakes on the Enriquillo fault and pointed out that a repeat of one of these earthquakes could cost hundreds of thousands of lives.

The government officials then asked: "When will the next earthquake strike our city?" The scientific response spoke to the uncertainty of earthquake science. We can point out the danger areas, but we cannot predict the time of the next Big One. It could strike tomorrow or a century from now. The Haitian government took this inability to provide the date of the next earthquake as a reason to place other pressing problems ahead of earthquake protection, as described in the Preface.

The Haitian government did take some action. The Civil Protection Agency of Haiti began a program of earthquake awareness. A meeting to discuss the first draft of a pamphlet on earthquake hazards to distribute to the local population was in progress when the earthquake struck. Many people at the meeting were killed, including the engineer in charge of Haiti's earthquake hazard program.

Haiti has major problems that need to be addressed today: hurricanes, disease, corruption, the influx of thousands of unemployed Haitians to the capital city looking for work, a decrepit building stock, and virtually no social services. Transparency International ranks the Corruption Perception Index for Haiti at 19, the worst in the Greater Antilles.

It is hard to blame Rene Preval, the president of Haiti, if he gambled and did not start a major program of strengthening buildings against earthquakes. His government didn't have the money. As it turned out, the president lost his gamble because the earthquake occurred while he was still in office. But he could not predict that at the time, and neither could the experts who had visited his government officials before the earthquake.

Several years after the Léogâne earthquake, the presidential palace and major government buildings have still not been repaired, and thousands of Haitians are still living in refugee camps. The Haitian economy has not recovered.

Haitian government officials asked the obvious question, but scientists could not give them the answer they needed. Since they didn't have the money to do on their own what needed to be done in several fields, not just earthquakes, can you blame them?

JAMAICA

West of the Tiburon Peninsula, the plate-boundary Enriquillo strike-slip fault continues westward beneath the Caribbean Sea as the

Plantain Garden fault (PGF, Figure 24.1). This fault comes ashore in Jamaica, where it bends to the northwest near the capital city of Kingston. The change in orientation means that the strike-slip Plantain Garden fault changes to a fault that squeezes the crust, uplifting it to form the island of Jamaica.

Jamaica was discovered by Columbus in 1494 and claimed by Spain. The English invaded the island in 1655 and took charge, following a treaty with Spain in 1670. The main city at that time was Port Royal, built on the Palisadoes sand spit that partially encloses the main harbor on the south. The English constructed several forts but turned over the policing of Port Royal to pirates who preyed on the Spanish galleon trade. Although the city, with a population of 6500, was one of the two most prosperous English-speaking towns in the Western Hemisphere, it was essentially lawless, known for drunkenness and prostitution, and as a lair for pirates. Crime was rampant.

In the late morning of June 7, 1692, disaster struck in the form of an earthquake that essentially destroyed Port Royal. The Palisadoes sand spit on which the town had been built suddenly liquefied, and most of the northern part of the city slid or subsided into the sea. The earthquake was followed by a tsunami. Between 1000 and 3000 people, nearly half the population, lost their lives, and an additional 2000 people died from disease that spread throughout the region shortly afterwards. The Reverend Cotton Mather of Boston, Massachusetts, preached that Port Royal's fate was caused by the sins of its citizens. The town never recovered; its population today is less than 2000.

Jamaicans feared returning to Port Royal, and so, in 1692, after the Port Royal earthquake, they established the city of Kingston on the north side of the harbor. Kingston became the seat of government, and its population gradually grew, particularly in the twentieth century. But it, too, was struck by a devastating earthquake of magnitude 6.5 on January 14, 1907. Every building in Kingston was damaged, and 85% of the buildings were destroyed. Between 800 and 1000 people died out of a population estimated as 50,000–55,000. This earthquake was also accompanied by a tsunami with maximum wave heights greater than seven feet (2 m).

Since the 1907 earthquake, Kingston has continued to grow in population. In 1921, 95,000 people lived in greater Kingston. The population was 338,000 in 1955, and it is almost one million today. About one in three Jamaicans lives in or close to Kingston.

Because of the growth in population, a repeat of the 1692 or 1907 earthquake would be catastrophic. Yet very little is known about the earthquake faults that threaten the city. It is not even clear that the two large historical earthquakes were on the Plantain Garden fault. In addition, there are other faults in the Blue Mountains northeast of Kingston and along the south coast of Jamaica west of Kingston that are clearly hazards but are poorly known. Some scientists at the University of the West Indies in a suburb of Kingston are concerned about the problem, but Jamaica is a poor country, and government funding is limited. Jamaican earthquake scientists have dealt with this problem by collaborating with scientists from the developed world, but there is no major government-sponsored program to evaluate Jamaica's faults or to upgrade building codes against strong ground motion. Transparency International gives Jamaica a Corruption Perception Index ranking of 38, considerably higher than Haiti, but low in comparison to developed countries.

GUANTÁNAMO BAY, CUBA (YEATS, 2014)

The north boundary of the Gonâve tectonic plate, like the south boundary, is a strike-slip fault called the Oriente fault, across which North America tends to move westward with respect to the Gonâve tectonic plate. The plate boundary is onshore in the Cibao Valley in the heavily populated northern Dominican Republic (DR, Figure 24.1), which has been struck by several large earthquakes since the mid-1800s. The fault skims the northern coast of Haiti and extends west to the linear offshore Oriente Deep just south of eastern Cuba, which gives the fault its name. Among the places at risk are the cities of Guantánamo (G, Figure 24.1) and Santiago de Cuba (SC, Figure 24.1) and the Sierra Maestra, birthplace of the Fidel Castro revolution that overthrew the dictatorship of Fulgencio Batista.

In terms of social services, Cuba is better off than Haiti or Jamaica, and Transparency International gives it a Corruption Perception Index of 48, the highest in the Caribbean. But most Cubans probably have no idea that part of their island is an earthquake hazard zone. The fault is entirely offshore south of Cuba, and it is identified mainly by earthquakes (Figure 24.1). Among these earthquakes are moderate-size temblors in 1852 and 1887, the last between northern Haiti and the eastern tip of Cuba. An earthquake on the Oriente fault west of Cuba in 1992 may be part of this westward-propagating sequence, but if it is, there is an unfilled seismic gap opposite Santiago de Cuba and Guantánamo. An offshore earthquake of magnitude 6 on September 15, 2011 produced moderate shaking in the province of Granma in southeastern Cuba.

The United States has a military prison at Guantánamo Bay that has been in the news because of prisoners accused of terrorism that are being held there. I have never seen any mention by the media of the proximity of the prison to the Oriente fault, but there is a real possibility that the area might be shaken violently by a major earthquake. Although President Obama has advocated closing the prison, Congress has not allowed him to do so. For many decades, the United States did not have normal relations with Cuba, so the significant contribution the Americans could make toward alleviating earthquake hazards at Guantánamo has not been considered. Now that normal relations are being restored, an American contribution could include upgrading the network of the Cuban Seismological Service, measuring tectonic displacements in far eastern Cuba using GPS, and reviewing building codes in eastern Cuba to ensure they are adequate to protect the population against earthquakes, including people at the US military base.

BOX 24.1   **Large cities in poor seismogenic countries**

The Haiti earthquake of 2010 illustrates the problem faced by poor countries when they are struck by massive earthquakes. These countries simply do not have the resources to respond to such an earthquake or even to plan for one. Richer nations have contributed to recovery in Port-au-Prince, but it is not enough. The Haitian economy is still in shambles and is likely to be for decades to come. The same thing was true for Managua, the capital of Nicaragua, destroyed by a moderate-size earthquake of magnitude 6.2 in 1972. That earthquake destabilized Nicaragua and was followed afterwards by the fall of the government of long-term dictator Anastasio Somoza Debayle and by decades of civil unrest between the Sandinistas and Contras during the 1980s, described in Chapter 26. Other Central American countries are also unable to prepare their major cities for the next earthquake.

In the Caucasus, the former Soviet republics of Armenia, Georgia, and Azerbaijan contain active faults. In the waning days of the USSR, Armenia was struck in 1988 by an earthquake of magnitude 6.8 that destroyed the city of Spitak, including taking the lives of 25,000 to 50,000 people, including most of its elementary school children, when school buildings collapsed. A major reason for the high losses was the poor quality of Soviet building construction during the era of Leonid Brezhnev, including school buildings. The Soviet Union was about to collapse, so Armenia was left on its own to cope with reconstruction after the earthquake.

Farther east, Turkmenistan is crossed by the Kopeh Dagh fault, which sustained a major earthquake that struck its capital of Ashkhabad (now Asgabat) in 1948. Turkmenistan had a different problem. In this case, Turkmenistan was a Soviet Socialist Republic, but for political reasons news of the earthquake was not made public by the Soviet media. The earthquake had a magnitude of 7.3, and most recent estimates of losses of life are as high as 176,000.

25 Mexico City: bowl of jello inherited from the Aztecs

INTRODUCTION AND A BRIEF HISTORY

Hernán Cortés and his small band of Spanish soldiers arrived in the Valley of Mexico on November 8, 1519, after a long march from the Gulf of Mexico. What they saw was unreal, expressed by some members of the expedition as like a dream. They and their Mexican allies looked out across Lake Texcoco to an island containing a beautiful city, with tall buildings, stone monuments, wide streets and canals, and long aqueducts, connected to the mainland by causeways (Figure 25.1). The Aztec city of Tenochtitlán had a population of more than 200,000, although some estimates of the number of inhabitants are as high as 400,000, which would have made it the largest city in the Americas. Either estimate made the city larger than any city in the native Spain of the invaders, and, for that matter, larger than any contemporary European city. The Aztec empire was called the Triple Alliance because it was a union of three separate city-states: Tenochtitlán, Tlacopan, and Texcoco. The empire, dominated by the Aztec/Mexica city-state of Tenochtitlán, stretched from the Pacific to the Gulf coast and extended south toward the Yucatán Peninsula, location of the older Mayan and Olmec civilizations.

The Spanish expedition was greeted with great respect by the Aztec emperor, Moctezuma, and his entourage. An Aztec legend held that at about this time, Tenochtitlán was supposed to be visited by Quetzalcóatl, the god of the East. Moctezuma concluded that the Spanish party, with their light skin, long beards, and short hair, having arrived on the east coast on tall sailing ships perceived by the Aztecs as floating towers, fulfilled that prophesy.

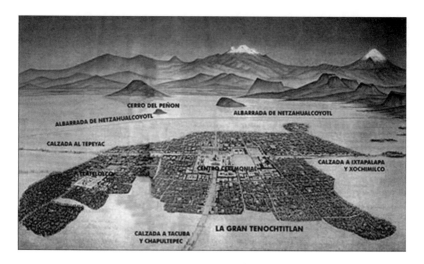

FIGURE 25.1 Artist's impression of the Aztec city of Tenochtitlán, as it would have appeared to the Spanish soldiers of Hernán Cortés. The city was situated on an island in Lake Texcoco connected to the mainland by causeways. View is toward Ixtaccíhuatl (left) and Popocatépetl volcanoes. After the Aztecs were defeated, construction was plagued by subsidence of water-saturated lake deposits. These deposits result in the "bowl of jello" effect, accentuating strong ground motion from earthquakes, including those on the subduction zone. Source: National Museum of Anthropology, Mexico City, painted by Atl in 1930.

However, it soon became apparent that the two cultures were not going to get along. The Spanish were shocked by the Aztec practice of human sacrifice, and the Aztecs were insulted by the insistence of the Spanish that the accepted religion would be Christian, not the Aztec god Huitzilopochtli.

The Spanish were greatly outnumbered, and they would have been overwhelmed quickly except that other Mexican tribes hated being lorded over by the Aztecs. The Aztec, or Mexica, people, originally from northern Mexico, were the most recent dominant group, and the Aztec empire was the largest in North America up to that time. The Aztec empire had been founded in AD 1325 and had expanded at the expense of older cultures that chafed under ruthless Aztec rule. Furthermore, Moctezuma's deference to the Spanish invaders was regarded as a sign of weakness by his subjects and by other tribes, and he lost favor. He died the following year, in 1520, and was replaced by a young cousin, Cuauhtémoc. But the new king was overthrown in 1521, and the Spanish took over the empire.

Were the Aztecs aware of an earthquake threat to Tenochtitlán? They did record earthquakes, as described in 2004 by Robert L. Kovach of Stanford University. The Aztec symbol for earthquake is *ollin* or *tlalli*. Several earthquakes in the Valley of Mexico were documented in the fifteenth century, before the arrival of Cortés. One of the Aztec gods was Tepeyollotl, the god of earthquakes, echoes, darkened caves, and jaguars! One Aztec legend held that the world would end with great earthquakes and the destruction of the sun.

The sun god, Tonatiuh, was also related to earthquakes. In addition, the Aztec sunstone was thought to be able to *predict* earthquakes. Unfortunately, there is no record of how they did it, or if they were ever successful.

By 1525, after the Spanish had taken over, the population of Tenochtitlán had dropped from more than 200,000 to about 30,000. The reason for this was the introduction of smallpox and other European diseases for which the Aztecs and their conquered tribes had no resistance. The city on Lake Texcoco was renamed Ciudad de México (Town of the Mexica), or Mexico City, by the Spanish, although its population remained predominantly Aztec and other indigenous tribal people. Only about 2000 residents were of European background. The city became the capital of the Viceroyalty of New Spain, and remained so for the next three centuries.

As the city expanded, it became apparent that it had a problem with flooding due to the rise in the level of Lake Texcoco during summer rains. In 1589, the Spanish king sent a German engineer, Heinrich Martin, who began to drain Lake Texcoco in 1629 through a new outlet, put in place in 1608. The problem with this solution was that large buildings began to sink into the soft lake sediments. This was recognized even during the Aztec period because the largest Aztec structures in Tenochtitlán devoted to the gods Huitzilopochtli and Tlaloc (Templo Mayor in Spanish) began to sink. In 1968, I visited one of the world's most revered shrines, the Basilica of Our Lady of Guadalupe, now the Antigua Basílica since a new church was built next to it between 1974 and 1976. The Antigua Basílica showed the effects of subsidence because of the soft lake sediments, and this was a major reason for building the new church.

In estimating the population of Mexico City today, it is necessary to ask: population of what? The Distrito Federal, the equivalent of the District of Columbia in the United States, had a population of nearly nine million in 2012. However, if the suburbs are included, the metropolitan region has a population of 18 million. Finally, it is estimated that 32 million people live in areas underlain by soft silt from the former Lake Texcoco. However, it will be shown below that, although subsidence of lake deposits is a major problem for Mexico City, it is not the worst problem to face the Valley of Mexico.

TRANS-MEXICAN VOLCANIC BELT

The Valley of Mexico, including Mexico City, is part of a volcanic–tectonic region called the Trans-Mexican Volcanic Belt (Figure 25.2). This region extends east–west, at a 15-degree map angle to the subduction zone between the North American continent and oceanic crust of the Cocos tectonic plate. The western end of the zone of active volcanoes is near the Pacific coast, east of the mouth of the Gulf of California, and the eastern end is along the shore of the Bay of Campeche in the Gulf of Mexico. Seismicity in the Trans-Mexican Volcanic Belt is much lower that it is on the subduction zone to the south, and active faults have slow slip rates. Nonetheless, the volcanic belt has been struck by earthquakes, with the largest the Acambay earthquake of magnitude 6.9 in November, 1912, which was accompanied by surface rupture. The Jalapa earthquake of magnitude 6.5 in January 1920 took more than 1500 lives, many from mudflows and debris flows triggered by the earthquake.

The Valley of Mexico owes its origin to faulting and volcanic activity in the Trans-Mexican Volcanic Belt (Figure 25.2). Mexico City

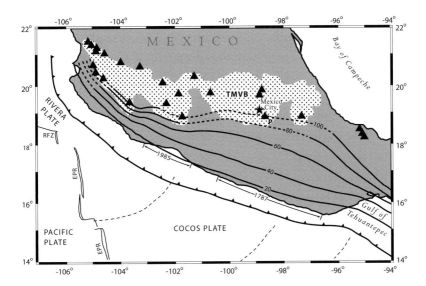

FIGURE 25.2 Location of Mexico City within the Trans-Mexican Volcanic Belt (TMVB). Filled triangles: active volcanoes; P, Popocatépetl; EPR, East Pacific Rise sea-floor spreading center; RFZ, Rivera fracture zone. Dates and arrows show those parts of the subduction zone that ruptured in 1985 and in the larger San Sixto superquake of 1787 in the states of Guerrero and Oaxaca. Contours in kilometers are on the Mexican subduction zone. Source: Yeats (2012, Fig. 4.18) and Andreani *et al.* (2008), used with permission.

is bounded on the east by two volcanoes, Popocatépetl (P, Figure 25.2), which is highly active, and Ixtaccíhuatl (Figure 25.1), which is much less active (both located east of Mexico City, marked on Figure 25.2 by a star). The highest mountain in Mexico is Pico de Orizaba (Citlaltépetl), which rises 18,491 feet (5636 m) above sea level, the third highest mountain in North America. Pico de Orizaba last erupted in 1846 and is currently listed as potentially active, with explosive activity and lava flows. The second highest mountain in Mexico is Popocatépetl, at 17,749 feet (5410 m), which I climbed in 1978 with my friend Chuck Denham. The climb started as a glacier ascent, but as we approached the summit crater, we became aware of acrid fumes of sulfur, and we could see columns of gas and steam emerging from the crater floor. Popo is currently in an active stage, and it is a threat to Mexico City. Recently, the mountain has produced ash plumes and volcanic bombs, and incandescent volcanic material has sometimes been visible at night from Mexico City.

At the eastern end of the Trans-Mexican Volcanic Belt, transitional to the volcanic arc of Central America, is El Chichón, a relatively insignificant-appearing mountain (chichón means bump) that had a massive eruption in 1982 that killed at least 1900 people, completely destroyed a village, and ejected enough sulfur dioxide and particulate matter into the uppermost atmosphere that it temporarily cooled the climate of the Northern Hemisphere. The eruption was seven times greater than the eruption of Mt. St. Helens in 1980.

MEXICAN SUBDUCTION ZONE AND THE 1985 MICHOACÁN EARTHQUAKE IN MEXICO CITY

The subduction zone on the southern coast of Mexico is marked by the highest seismicity in the country. This region has been struck frequently by earthquakes, with 42 earthquakes larger than magnitude 7 in the twentieth century. The Jalisco earthquake of magnitude 8 struck in June 1932 near the western end of the subduction zone, resulting in at least 400 dead. This earthquake caused several buildings in Mexico City to topple over, and there were cracks in the streets, but because the Jalisco earthquake was far away, the damage was not enough to trigger a major alarm in the capital city.

None of the earthquakes on the subduction zone had caused great losses in Mexico City until September 19, 1985, when an earthquake of magnitude 8.1 struck the coastal region in the State of Michoacán and the western edge of the adjacent State of Guerrero, south of Mexico City (Figure 25.2). There was significant damage along the Mexican coast, and a tsunami with wave heights up to ten feet (3 m) was generated. But in Mexico City, more than 200 miles (320 km) away, the damage was catastrophic. A widely cited death toll was 10,000 people, but this figure was not trusted by the public, and many sources estimated that the figure was closer to 40,000. Official government estimates were that 250,000 people had lost their homes, but these figures also were not believed by the general public, and a figure of 700,000 was widely accepted.

Why was Mexico City hit so hard, since it was located so far from the epicenter? The reason is that the city has expanded on top of the former bed of Lake Texcoco, which had been drained, but still has water-saturated lake silts beneath the surface. To visualize what happened, try this experiment. Imagine a bowl of jello sitting on a table. To represent buildings, stack some small blocks on top of the jello, and another stack of blocks on the table top. Now shake the table. The blocks on the bowl of jello would fall down, whereas the blocks on the table would not. This is essentially what happened in Mexico City, where 32 million people live on top of the silts of Lake Texcoco. These silt deposits behave like jello when they are shaken. Seismic waves reaching the city from the coast slow down when they reach the lake deposits. At the same time, the waves become more violent, like the response of the bowl of jello.

Another factor increasing the damage was the duration of strong ground shaking: three to four minutes. In Chapter 3, I recounted my own experience in Mexico City during an earthquake where the duration of strong shaking was only 15 seconds. My OSU colleague, Chris Goldfinger, experienced the 2011 Tohoku-oki earthquake with strong shaking lasting several minutes, which is comparable to that experienced in the Mexico City earthquake. A building might survive strong shaking of 10–30 seconds, but could be shaken apart by strong ground motion lasting several minutes.

Mexico City has many tall buildings, but the buildings that fared the worst were those of 6–15 stories. Engineers call this the *tuning fork problem*. The buildings resonated (vibrated) at the same frequency as the earthquake waves passing through soft lake sediments, thereby reinforcing these waves, and causing the buildings to shake violently and collapse. (Think of the fabled operatic soprano who belts out a loud, high-pitched aria, which causes glassware to break.) In contrast, taller buildings performed relatively well. For example, the 44-story Torre Latinoamericana (Latin American Tower) was virtually undamaged. Contributing to its survival was the fact that it was constructed on pilings that had been driven 100 feet into the lake silts beneath the surface.

Another major factor in the large loss of life was the response of the Mexican government and its president, Miguel de la Madrid of the ruling PRI (Partido Revolucionario Institucional). The president imposed a news blackout. The government refused aid from other countries, including the United States, despite the fact that Mexico was in the midst of a financial crisis. There were no seismic building codes prior to 1957, and buildings constructed between 1957 and 1976 were particularly hard hit. The government response preferentially favored companies with strong political ties to PRI, even assigning soldiers to help remove equipment from damaged factories before removing the dead bodies of victims. The government's figures for losses of life and people made homeless were not trusted by the public. De la Madrid himself was often not to be found, although he did declare a three-day period of mourning for the victims of the earthquake. When he finally did make a public appearance, he was heckled, in contrast to the highly respectful treatment by the public of his PRI predecessors. The collapse of recently constructed buildings led to charges of corruption and incompetence. This led to major unrest among the affected population of the city, resulting in the growth of opposition parties and, eventually, in 2000, a change of government to the Partido Acción Nacional, or PAN.

There were also signs of progress. Mexican citizens, operating as neighborhoods (*colonias*), volunteered unselfishly for search and rescue, often without regard for their own safety. The government sponsors evacuation drills on September 19 of each year, the anniversary of the great 1985 earthquake. Seismologists at UNAM, the national university, recognized that it took time for the earthquake signal recorded on coastal seismographs to reach Mexico City after the mainshock on the offshore subduction zone. In the 1990s, they established sensors along the subduction zone (Sistema de Alerta

Sísmica) to take advantage of the delay between the mainshock and the arrival of strong-motion seismic waves at Mexico City. This enabled them to send an electronic warning to Mexico City in time to automatically shut down essential services such as hospital operating rooms, train lines, and subways at the sign of a large earthquake at the coast.

I call this "predicting an earthquake after it happens." It has been highly successful in Mexico.

A spokesman for the National Center for Disaster Prevention points out that building codes have been upgraded since the 1985 earthquake, including a major strengthening in 2004 against earthquakes. It remains to be seen if the corruption problem has been solved. Transparency International ranks Mexico's Corruption Perception Index at 34, a relatively low number that may reflect the Mexican government's response to drug cartels as well as to enforcement of building codes.

WHAT ABOUT THE NEXT GREAT EARTHQUAKE?

If the 1985 earthquake relieved the strain on the subduction zone south of Mexico City, is there still a problem? Sadly, there is. The 1985 earthquake filled a seismic gap (marking a long period without large earthquakes) in Michoacán and westernmost Guerrero states (Figure 25.2). Unfortunately, there is another gap to the east in Guerrero State, due south of Mexico City, called the Guerrero Gap. This part of the subduction zone is being intensely monitored by Mexican seismologists, who have observed evidence of movement on the subduction zone not accompanied by a major earthquake, called *slow slip events*, resembling features observed on subduction zones in the Pacific Northwest and in southwest Japan. An earthquake of magnitude 7.2 struck the edge of this seismic gap on April 18, 2014, but this was not the Big One that Mexico is waiting for.

The 1985 earthquake was larger than others along the subduction zone, leading planners prior to the earthquake to conclude erroneously that the largest earthquake they should plan for would have a magnitude of 8 or less. Japanese planners made the same mistake in using magnitude estimates of 8.4 or less for earthquakes on the northeast Japan subduction zone as the largest earthquake that they needed to plan for. But the March 2011 superquake had a magnitude of 9, large enough to produce a tsunami that took nearly 16,000 lives and caused the destruction of the Fukushima Dai-ichi nuclear power plant. If they had looked back 1000 years rather than only a century, they would have identified an older earthquake that was at least as large as the 2011 earthquake (see Chapter 9 on Tokyo and the Kansai).

Mexican scientists have looked back at the pre-twentieth-century earthquake history of their subduction zone. This is not easy to do because of the scarcity of settlements on the Pacific coast in the nineteenth century and earlier. But an earthquake on the coast east of the Guerrero seismic gap on March 28, 1787, called the San Sixto (Guerrero-Oaxaca) earthquake (located on Figure 25.2), is now regarded as the largest earthquake in Mexico's recorded history. This earthquake was accompanied by a tsunami with wave heights of 30–40 feet (9–12 m). Records of earthquake intensity suggest that this earthquake ruptured along a section of the subduction zone 300 miles (480 km) long, twice as long as the rupture on the subduction zone in 1985 (Figure 25.2). Estimates of the magnitude of this earthquake are as high as magnitude 8.6–8.7, a superquake, like the March 2011 Tohoku-oki earthquake off northeastern Japan.

So there is little doubt that the preparation of Mexico City against the next great earthquake will be tested. Will the inventory of buildings on top of the saturated silts of Lake Texcoco be replaced by buildings able to withstand strong ground shaking, as the Torre Latinoamericana did in 1985? Will the government corruption exposed in the 1985 earthquake be replaced by honest builders and incorruptible building inspectors, a problem shared by many other megacities worldwide?

26 Central America and the earthquake that brought down a dictator

INTRODUCTION

Central America consists of seven small countries between Mexico and Colombia, most of which are afflicted by earthquakes and volcanic eruptions. These include Guatemala, Belize, Honduras, El Salvador, Nicaragua, Costa Rica, and Panama. All were formerly ruled by Spain except for Belize, which was formerly British Honduras and is the only country not in danger from earthquakes or volcanoes. Mexico tried unsuccessfully to annex the five countries of Guatemala, Honduras, El Salvador, Nicaragua, and Costa Rica after they declared their independence from Spain, but the Central American countries, led by Guatemala, broke away and formed the United Provinces of Central America, which itself broke up into separate countries in 1838. Panama was formerly part of Colombia, but achieved its independence in 1903 with the backing of the United States because it was the site of the future Panama Canal. Most of these countries have had difficult, even violent, political histories (LaFeber, 1983), with the exception of Costa Rica, which became a democracy and did away with its army. Costa Rica is not a wealthy country, but it is a stable one.

The five countries from Guatemala to Costa Rica contain one of the densest concentrations of active volcanoes in the world (inset, Figure 26.1). In addition, the Middle America subduction zone has generated many earthquakes, but this subduction zone is very different from the Mexican subduction zone to the north. Both subduction zones involve the Cocos plate, but the dip of the oceanic Cocos plate opposite Central America is much steeper than it is in Mexico. Because of this steep dip, earthquakes on this subduction zone are offshore and have not produced a hazard as high as earthquakes on the low-dipping Mexican subduction zone. People have lost their lives in Middle America subduction zone earthquakes, but the magnitudes are smaller, and none have been as large as magnitude 8.

A greater hazard is earthquakes in continental crust, generated on faults close to Central American volcanoes. In Guatemala, El Salvador, and Nicaragua, many of these faults tend to be strike-slip, in which crust closer to the ocean moves northwest during an earthquake, parallel with the coastline, with respect to the crust in the interior. These faults are relatively short, and some have only been

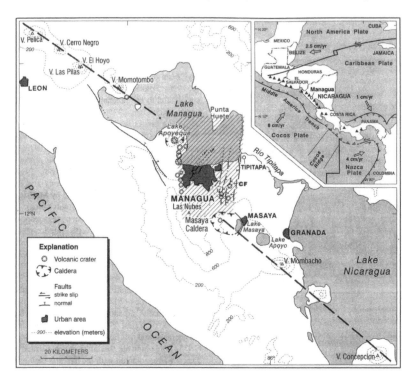

FIGURE 26.1 Location of Managua, Nicaragua, on the shore of Lake Managua, with active faults that were the sources of destructive earthquakes in 1931 and 1972. CF, Cofradia fault. Active volcanoes shown as open triangles and crater symbols in main map; solid triangles in inset map. Source: Yeats (2012, Fig. 4.17) and Cowan *et al.* (2002).

identified by earthquake geologists in recent years. These are the faults that generated the earthquakes that struck Managua, Nicaragua (Figure 26.1), and San Salvador, the capital of El Salvador, as discussed below. Another set of faults is part of the plate boundary between the North American and Caribbean plates, also marking the boundary between the Mexican and Middle American subduction zones. These plate-boundary faults have been the source of major earthquakes in Guatemala.

MANAGUA, NICARAGUA, THE UNITED STATES, AND SOMOZA

Nicaragua's problem in the mid-nineteenth century was that it is too close to the United States, which was embarking on its expansionist goal of Manifest Destiny. The Republic of Texas, established in 1836 in territory formerly part of Mexico, was annexed in 1845 as the 28th state, leading to a war with Mexico in 1846–1848. This war resulted in the addition to the United States of a large part of northern Mexico, including New Mexico and California, in addition to Mexico's being forced to recognize Texas as part of the United States. This caused controversy in Congress because Texas had been admitted as a slave-owning state, and Northern legislators argued that this changed the balance of power in Congress between slave-owning and slave-free states. The small countries of Central America became pawns in this conflict between Northern and Southern states.

Stepping into the fray was an American adventurer, William Walker, who was the best-known mercenary soldier to attempt to take over various territories in Latin America. Walker was called a *filibuster*, derived from the Spanish *filibustero* and the pirate term *freebooter*, not to be confused with the recently modified practice of the US Senate to require 60 votes to pass a bill and avoid a "filibuster" that would delay action on controversial nominations and legislation proposed by the president. Walker began with an attempt in 1853 to establish an independent country in Baja California and Sonora in northwestern Mexico, which was initially successful but soon failed.

This was followed in 1855 with an invasion of Nicaragua by Walker and an armed band of Americans during a time of civil war in Nicaragua. Walker set himself up as president, re-established slavery so that Nicaragua could be admitted to the Union as a slave state, and made English one of the official languages. However, Walker was unpopular in Nicaragua. After being forced out of that country, he was captured and executed in Honduras in 1860, and his filibuster failed.

Although this adventure marked the end of the southward territorial expansion of the United States in the North American continent, it was only the beginning of its involvement in Central American politics (LaFeber, 1983). Rather than acquisition of territory, the American strategy was to dominate the small Central American countries through acquisition of companies producing raw materials, especially the United Fruit Company, leading to the term "banana republic" to describe nations of the region. The Americans used their influence on imports and exports to the United States and on financing through American banks to dominate the political life of Central America, including building railroads. The production of bananas, coffee, and tropical wood led to the concentration of land in the hands of a small number of wealthy families (oligarchs), at the expense of the lower classes who worked fields they did not own for essentially slave wages. The United States, formed by a revolution, found itself in the unfortunate position of supporting those rich families who were producing the agricultural products and reaping the benefits at the expense of the general population. This required American support of local armies that enforced these regulations ruthlessly, even into the late twentieth century. These armies were trained and supported financially by the United States.

After Walker's overthrow, the Conservative Party was in power in Nicaragua from 1857 to 1893. In 1893, José Santos Zelaya came to power in Nicaragua during a time when the United States was campaigning to reduce the influence of European powers, especially the United Kingdom, in Central America. By 1903, Zelaya was at odds with the dictator of neighboring Guatemala. Warfare between these countries led to the involvement of the United States during the presidency of Theodore Roosevelt. Despite the appearance of an American warship, Zelaya continued fighting, particularly in Honduras, opposing the increased involvement of the United States. A revolution began in Nicaragua, resulting in the loss of the British-controlled Miskito Coast to revolutionaries and the execution by Zelaya of two Americans in the town of Bluefields on the Miskito Coast.

The United States dispatched 2600 Marines to Nicaragua in 1912 to stop the fighting. US Marines would remain in Nicaragua for the next 21 years until 1933, leaving in the early years of the Great Depression. Much of the 1920s was marked by civil war between the Conservatives and the Liberals, and by the role played by Augusto Sandino, who opposed North American capitalist exploitation and the meddling of the American military in Nicaragua's internal affairs. Despite Sandino's efforts, including guerrilla warfare, the Americans supervised the formation of a National Guard, which would play a major role in the rise to power of the Somoza family. Sandino had disappeared into the mountains in 1927, and he became a popular, charismatic leader among the Nicaraguan *campesinos*, who had been ignored in previous campaigns by the United States to gain influence among the oligarchs. In 1933, after the end of occupation by the unpopular US Marines, Sandino reached an agreement with the ruling Liberal government. But the National Guard, under its commander Anastasio Somoza García, had Sandino assassinated in 1934. By 1936, the Somoza family, with US backing, was running the country and would do so for the next 43 years.

Somoza deposed the elected Liberal Party president, Juan Bautista Sacasa, and took over the country. The Somoza family controlled the ruling Nationalist Liberal Party, which controlled the legislature and the judiciary, meaning that the Somozas had dictatorial powers. A US-owned timber company paid Somoza to permit them to log much of the tropical forest. Pesticides, including DDT, banned in the United States, were widely used in Nicaragua. Nicaragua became the principal beef exporter to the United States, using slaughterhouses owned by the Somoza family, later headed by Anastasio Somoza Debayle, a family that became fabulously wealthy. Despite his record and his dictatorial rule, Somoza García was warmly received in Washington in 1939 by President Franklin Roosevelt, where he addressed a joint session of Congress and received $2 million in credits from American banks.

MANAGUA EARTHQUAKE OF 1972 AND THE END OF THE SOMOZA DICTATORSHIP

On December 23, 1972, an earthquake of magnitude 6.2 scored a direct hit on the million residents of Managua, the national capital. Five thousand people were killed, 20,000 injured, and more than 250,000 made homeless. The earthquake came as a complete surprise to the Somoza-dominated government, although it should not have. It had been preceded by an earthquake of magnitude 6 on March 31, 1931, which took 2000 lives.

Both earthquakes originated not from the subduction zone but from crustal faults close to the city (Figure 26.1; Cowan *et al.*, 2002). These earthquakes were much smaller than subduction zone earthquakes, but they caused major damage because they originated so close to the surface. The Somoza government was completely unaware of the hazard from these faults, as well as the hazard from active volcanoes in the same region.

The economy collapsed due to the 1972 earthquake. All four hospitals were unserviceable, and only 10% of the population of the city had water a week after the earthquake. Fires raged for several days because firefighting equipment was inoperable. Disaster assistance came from the United States, Mexico, and 25 other countries, but its distribution to earthquake survivors became a fiasco.

Somoza's family was now run by Anastasio Somoza Debayle after the assassination of his father, Anastasio Somoza Garciá, in 1956. Somoza family members plundered much of the foreign aid and enriched themselves at the expense of the victims of the earthquake. The rebuilding that did take place was on Somoza-owned land by Somoza-owned contractors, funded through Somoza's banks. This was widely reported, and opposition to the Somoza regime grew. The lower classes were hardest hit by the earthquake and the government response, and this led to growth of a revolutionary movement that had begun in the 1960s, called *Frente Sandinista de Liberación Nacional* (FSLN), or Sandinista National Liberation Front, named after Augusto Sandino, who had been murdered under the Somoza family's orders. Somoza's profiteering after the earthquake led to increased support for the FSLN, and President Jimmy Carter announced that the United States was withdrawing its support of the Somoza regime. In 1979, Somoza, together with the commanders of his National Guard, was forced to flee the country, to be replaced by the FSLN.

Nicaragua's problems with the US government were not over. The FSLN had been inspired by the Cuban revolution, and it regarded itself as a revolutionary socialist country. A provisional government by the FSLN, led by Daniel Ortega, took over in 1979 after the departure of Somoza, and Ortega won a national election in 1984. President Ronald Reagan became alarmed at the support provided to the FSLN by the Soviet Union, Cuba, and other Communist countries, including a massive build-up of Nicaragua's military. Nicaragua then became caught up in the Cold War. The United States began supporting a group called the Contras, including Somoza supporters in the former National Guard, who were provided military supplies illegally through the CIA. Fighting between the Sandinistas and the Contras continued until a cease-fire was established in 1988. Since then, elections have been more or less fair, and in 2006 Daniel Ortega was again elected president of Nicaragua.

It is not the purpose here to describe in detail the political history of Nicaragua from 1979 until today, except to point out the problems leading to Somoza's departure began with the Managua earthquake of 1972. Reconstruction of the city did not begin until 20 years after the earthquake, after the Sandinista–Contra fighting had ended. Managua, an earthquake time bomb, became an example of how a small country can become destabilized by a natural disaster when it is caught up in international politics. Other time bombs have the same potential: Tehran, Kabul, and Caracas, for example. Damascus and Aleppo in Syria are disasters waiting to happen.

EARTHQUAKES IN EL SALVADOR

El Salvador is much smaller than its neighbors, Nicaragua and Guatemala. It extends parallel to the Pacific coast and is at risk from earthquakes and tsunamis on the offshore subduction zone as well as earthquakes on crustal faults close to active volcanoes. The country has had a troubled political past (LaFeber, 1983), including authoritarian governments from the 1930s to the 1970s and a civil war from 1980 to a peace accord in 1992. The region is characterized by a high crime rate, leading to emigration of many Salvadorans. More than 25% of the Salvadoran population now lives in the United States.

The country has been struck by many earthquakes in its history, and several have caused damage to the national capital, San Salvador, with the current population of its metropolitan region of 2.3 million. Three recent earthquakes are mentioned here, with the first directly beneath San Salvador on October 10, 1986, with a magnitude of only 5.7. Despite its small size, the earthquake resulted in 1500 deaths and $1.5 billion in damages. A larger crustal earthquake of magnitude 6.6 struck the region on February 13, 2001, taking the lives of more than 300 people. A month earlier, an earthquake of magnitude 7.7 on the subduction zone took more than 900 lives, many from landslides accompanying the earthquake.

The larger earthquakes, in some cases accompanied by tsunamis, generally are not as destructive as crustal earthquakes such as the one in 1986, which took 1500 lives despite its small magnitude. The reason for the great loss of life is the absence of structures reinforced against strong shaking, together with the landslides that often accompany crustal earthquakes in this tropical, high-rainfall region.

The landslides themselves are aggravated by the extensive deforestation near population centers, including San Salvador. The pro-business government appears reluctant to approve zoning laws that would prevent development in unsafe parts of the city susceptible to landslides and strong ground motion. The national government will take modest steps to address the earthquake problem after the fact, but little or no action to prepare San Salvador and other cities against the next inevitable earthquake, despite the frequency of earthquakes throughout El Salvador's history.

THE MOTAGUA, GUATEMALA, EARTHQUAKE OF 1976

Guatemala contains a major plate boundary between the Caribbean plate on the south and the North American plate on the north (inset, Figure 26.1). In the Greater Antilles to the east, this plate boundary consists of a southern zone of faulting that was the source of devastating earthquakes in Haiti in 2010 and in Jamaica in 1692 and 1907 and a northern zone in the northern Dominican Republic extending west as the Oriente fault, offshore Guantánamo, Cuba (see Chapter 24). In the western Caribbean, these two plate boundaries merge into one and reach the Central American coastline near the international border between Honduras and Guatemala (inset, Figure 26.1). In Guatemala, the plate boundary is expressed as a zone of faulting up to 50 miles (80 km) across, with the most active structure the Motagua strike-slip fault. This fault roughly follows the Motagua River, which itself contains the major population centers of Guatemala, including Guatemala City.

The Motagua fault was the source of an earthquake of magnitude 7.5 on February 4, 1976. This earthquake produced major damage and loss of life in the valley of the Motagua River (Espinosa, 1976; Plafker, 1977). The death toll of 23,000 was large because the earthquake struck at 3 a.m., when most people were asleep. More than 1.2 million people were left homeless. The president of Guatemala, Kjell Eugenio Laugerud, organized a helicopter tour of the damaged area for ambassadors, which led to major international assistance. The USGS sent a team led by George Plafker to study the earthquake in detail. Plafker mapped the surface rupture, more than 142 miles (228 km) long, including rupture on a secondary fault called the Mixco fault, only six miles (10 km) west of the capital, Guatemala City. Plafker's team found that the earthquake was accompanied by more than 10,000 landslides, which accounted for much of the damage and loss of life. In fact, the first recorded disaster in Guatemala in 1541, with the loss of life originally thought to be the result of an earthquake, was later attributed to a type of landslide called a debris flow.

Because the earthquake damaged the economic heart of the country, its economic effects were devastating. This was complicated by the fact that Guatemala was engaged in a long-running civil war, with major players including the ruling white property owners supported by the army, trained by the United States, and the large Mayan population, forced to work for low wages at the same time they faced expropriation of their lands. The US-based United Fruit Company, which owned parts of the economy, including its railroads, in addition to fruit production, was part of the ruling establishment permitting the exploitation of the Mayas.

Although the civil war ended in 1996, it is unclear how much effect this history of civil unrest will have on preparation for the next plate-boundary earthquake. Guatemala's population has doubled since the 1976 earthquake, and the losses from the next earthquake could be much larger than the losses in 1976.

Earlier earthquakes in the region in the sixteenth century and in 1765 and 1773 might have occurred on the Motagua fault, although this has not yet been demonstrated. In addition, the Chixoy-Polochic fault, north of and parallel to the Motagua fault, was the source of an earthquake on July 22, 1816 that might have been larger than the 1976 earthquake. Archaeological evidence suggests still-earlier earthquakes in AD 950–1000 (Kovach, 2004), indicating that earthquake return times may be relatively short. Ancient Mayan inscriptions include pictographs of earthquakes.

LESSONS LEARNED

Central America consists of small countries with rapidly growing populations but without the financial resources of larger countries.

In addition, with the exception of Costa Rica, these countries lack the democratic traditions of the United States and Canada, resulting in civil unrest, dictatorships, and lack of preparation against future earthquakes. The democratic traditions of the United States have not translated to the lower classes of Central America. The United States has consistently favored the ruling oligarchs over the working poor, even sending in the military to help enforce the status quo. The Managua and San Salvador earthquakes were only moderate in size, and in the developed world they might have resulted in no deaths at all. Because of poor construction by corrupt contractors and poor preparation of their populations, the death tolls were large, and the ruling infrastructure did not appear to learn from the devastation they had just experienced. Many of the deaths were attributed to landslides; future development should take landslides into consideration, as is done in California.

In recognition of the corruption problem, Transparency International gives Nicaragua a CPI rating of 29 and Guatemala a rating of 33. Costa Rica, on the other hand, has a rating of 54.

PANAMA CANAL

This chapter closes with the example of Panama, which now is in charge of the Panama Canal. The Canal is being expanded to accommodate larger ships, and part of the environmental study required before construction of the expansion could begin was an investigation of earthquake hazard. More than a century ago, Panama had been selected over Nicaragua as the site of the Canal because it was believed at the time that Panama did not present a volcanic or earthquake hazard (McCullough, 1977). However, McCullough described damaging earthquakes during the more than 40 years that the Canal was being built. More recently, a research team headed by Paul Mann, then of the University of Texas at Austin, found evidence of active faults in Panama, and earthquake geologists from my consulting firm found evidence of active faulting at the Canal itself. One of these faults actually offsets the Spanish cobblestone trail (Camino de

Cruces) used to transport Inca gold across the Isthmus of Panama. This fault was the source of an earthquake on May 2, 1621, that heavily damaged Panama Viejo, the original Panama City.

The environmental analysis for the expansion of the Panama Canal should take earthquake hazards into consideration, although if it does, this will lead to higher reinsurance rates for the Canal expansion. However, I am concerned that this analysis may not extend to Panama City or even to those parts of the Panama Canal that are not being retrofitted for the expansion.

27 East African Rift Valley: a tale of two cities

INTRODUCTION

East Africa is tectonically coming apart. The ancient African continent, hundreds of millions of years old, is being broken up by faults accompanied by volcanoes and lava flows. Millions of years from now, if you think about the region's future as a geologist might, it will probably be a group of large islands that are actually small continents in the Indian Ocean, like the island of Madagascar. In fact, the northeastern part of Africa has already broken away from the rest of the continent to form the Arabian Peninsula. Arabia and Africa are now separated by the Red Sea, a newly formed ocean.

This chapter discusses two cities in the Rift Valley, Nairobi and Addis Ababa (Figure 27.1), both of which have a major earthquake hazard (Gouin, 1979); however, the hazard differs between the two cities.

NAIROBI

West and south of the Red Sea, the African continent appears to be still together, but it is cut by huge faults that form the African Rift Valleys. The eastern Rift Valley, called the Gregory Rift in Kenya, is bounded by great faults, and the center of the valley is downdropped below a high plateau, similar to the much larger continental plateau marking the tectonically more stable part of Africa to the west. Kenya's largest city, Nairobi, is on the plateau at an elevation of 5400 feet (1645 m), about the same elevation as Nairobi's sister city of Denver, Colorado. The high elevation means that even though Nairobi is on the Equator, it has a comfortable climate, accounting for its recent

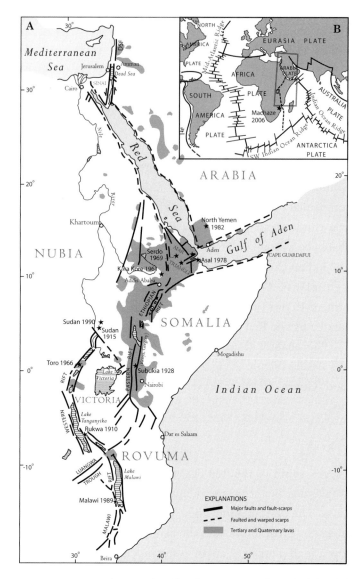

FIGURE 27.1 East African Rift Valley, locating Nairobi and Addis Ababa together with major earthquakes (filled stars and dates). The 1910 Rukwa earthquake was the largest historical earthquake in the Rift Valley, although the two Sudan earthquakes were also large. Subukia and Kara Kore earthquakes are discussed in the text. Red Sea and Gulf of Aden represent spreading centers marking the separation of African plates from the Arabian plate. Source: modified from Yeats (2012, Fig. 6.1) and Gouin (1979).

growth as a major commercial center, with a population greater than 3.1 million. By 2025, its population is expected to grow to five million.

Nairobi is a relatively new city, founded in 1899 as a rail depot between the cities of Mombasa, on the Kenyan coast, and Kampala in Uganda. In 1906, Nairobi's population was only 11,500. A year later, this small town was made the capital of British East Africa, in large part because its climate was agreeable to Kenya's English colonial rulers. Nairobi grew to its present population in the last few decades as a result of the large number of wealthy Kenyans and Europeans who established businesses there and built skyscrapers, and the migration to the city of poor Kenyans looking for work and to improve their lot in life. The influx of poor people, most from tribes living in small villages throughout Kenya, means that the city has spawned huge slums. The Kibera slum on the west side of Nairobi, with a population greater than 170,000, is the second largest slum in Africa.

The Rift Valleys are major tourist attractions because they have spectacular landscapes, and they are home to great numbers of large wild animals. They are still being formed as Africa breaks up, and the faults that bound the Rifts are subject to earthquakes, some with magnitudes greater than 7. The Rift Valleys stand out on a seismicity map (Figure 1.5) simply because they have experienced many small to moderate earthquakes. Some are felt in Nairobi, including an earthquake of magnitude 4.6 in April 2012 on a Rift Valley fault only 20 miles (32 km) to the west.

The largest Rift Valley earthquake in the past century was the 1910 Rukwa, Tanzania, earthquake of magnitude 7.4 in a rural region between Lake Tanganyika and Lake Malawi (Ambraseys, 1991). An earthquake of magnitude 7.2 struck near the city of Juba in South Sudan in May 1990, leaving 300,000 homeless. An earthquake closer to Nairobi, the Subukia earthquake of January 6, 1928, of magnitude 6.9 (located on Figure 27.1), was accompanied by surface rupture on the Gregory Rift west of the city.

None of these earthquakes raised an alarm, and it is not known if the Subukia earthquake was accompanied by any loss of life.

The reason for this is that East Africa was much more thinly populated, and people survived on a subsistence tribal economy, living in simple huts that were unlikely to kill their inhabitants when they collapsed in a magnitude 7 earthquake. An illustration of this is the February 2006 Machaze, Mozambique, earthquake of magnitude 7, which, despite its size, killed only a few people because it struck a sparsely populated region of small villages with simple dwellings. That was the case at the time of the Subukia earthquake in Kenya, when Nairobi, then as now the largest city of the region, had a population of only 31,000.

The situation is vastly different today because of the growth of East African cities.

The explosive growth of Nairobi is a recipe for disaster. Kenya, especially Nairobi, lacks the lesson of a devastating earthquake in its short recorded history killing tens of thousands of people, and therefore earthquakes, not being part of Nairobi's culture, are not considered as a major problem. Large earthquakes have struck elsewhere in the Rift Valleys, and in my opinion, these earthquakes are evidence that Nairobi must plan for an earthquake with a magnitude at least as large as the magnitude 7.4 Rukwa earthquake of 1910.

It is possible that the major skyscrapers of Nairobi have been built against the hazard of earthquake shaking because this would be required to obtain insurance against earthquakes, but it is unlikely that seismically resistant building codes protect the homes of the 3.1 million people of Nairobi, even those who are well off. Protection of housing against shaking is especially a problem in the slums of Nairobi. The people living there lack most basic services, and it would be unlikely that slum housing has been built with earthquake shaking in mind. In addition, the slum dwellers would not be living in simple huts as their tribal grandparents did, and their houses, like those elsewhere in the developing world, would collapse and kill many tens of thousands.

An example of what could happen is provided by an earthquake in 1982 of magnitude 6 east of the Red Sea in the Arabian peninsula south of Sana'a, the capital of Yemen (located on Figure 27.1). This earthquake, studied by George Plafker of the USGS, struck just east of the Red Sea rift valley. Despite its low magnitude, the earthquake took the lives of 1700–2800 people.

ADDIS ABABA

The capital of Ethiopia, Addis Ababa, lies on the northern extension of the Gregory Rift, but at a place where the earthquake history is different. Like Nairobi, Addis Ababa has a large population, now more than 3.4 million, and it is growing rapidly. It lies on a large plateau at an elevation of 7500 feet (2300 m). The city is the headquarters of the African Union and of the United Nations Economic Commission for Africa. Like Nairobi, Addis Ababa is subject to earthquakes, but unlike Nairobi, engineers in Addis Ababa are more aware of their earthquake hazard than are those in Nairobi.

The reason is that they have had a lot more of them. Addis Ababa University has a Geophysical Observatory, headed for many years by Father Pierre Gouin, a Jesuit seismologist from Canada whose work contributed to this analysis, including Figure 27.1. In addition to establishing seismographs in Ethiopia, he also spent several years developing an understanding of earthquakes in Nairobi and adjacent areas.

In the twentieth century, Addis Ababa was affected by two major earthquakes. The first, of magnitude 6.75, struck in August, 1906, and was preceded by a large foreshock, and the second, the Kara Kore earthquake of magnitude 6.7, took place in mid-1961 (located on Figure 27.1). The Kara Kore earthquake, studied in detail by Father Gouin, differed from those in Kenya and to the south. The earthquakes farther south, including the Subukia and Rukwa earthquakes, consisted of a mainshock followed by aftershocks, typical of earthquakes around the world. Because the time of the mainshock of such an earthquake cannot be predicted, such earthquakes come as a complete surprise, resulting in losses of life.

In contrast, the Kara Kore earthquake was not a mainshock–aftershock sequence but a series of smaller earthquakes called an *earthquake swarm*, similar to the Haicheng earthquake in northeast China that had been successfully "predicted." More than 3500 earthquakes were recorded between the end of May and September 1961, with 350 earthquakes per day at the beginning of June. In an earthquake swarm, the largest earthquake is not necessarily the first one; it might occur during the middle of the sequence. An earthquake swarm gets people's attention and might cause panic, but it also causes them to take steps to protect themselves. As a result, there was little or no loss of life in Addis Ababa from the two largest earthquakes of the twentieth century.

Why is Addis Ababa so different from Nairobi since both are part of the East African Rift Valley? In Kenya, individual faults have much greater displacements; in Ethiopia, there are many more faults, and the displacement on individual faults is much less than in Kenya. Addis Ababa itself has several mapped faults, and one of them is the site of a hot spring. The Ethiopian landscape has a much greater number of volcanoes and lava flows, and the high geothermal heat represented by hot springs means that the crust is weaker than it is to the south. Weak crust should not be able to store as much elastic strain as the stronger crust near the Rift Valleys to the south, so it should not generate such large earthquakes.

Does this mean that Addis Ababa is safer against earthquakes than Nairobi? Not necessarily. The epicenters of the two largest twentieth-century earthquakes affecting Addis Ababa were nearly 60 miles (100 km) away from the capital city. The damage might have been much greater if the earthquakes had been closer to the city. The analogy for an earthquake in weak crust is to compare a rotten log with a steel pipe. A shock to the rotten log dies out close to the source of the shock, whereas a shock to a steel pipe is still powerful far from the source. One goes *thunk*, and one goes *clang*.

Also, a large earthquake in the middle of an earthquake swarm might be anticipated, but if buildings were not reinforced against earthquakes, many people might lose their lives. An illustration of this was the L'Aquila, Italy, earthquake of April 6, 2009, of magnitude 6.3. The mainshock occurred in the middle of an earthquake swarm that had been going on for more than three months, causing great alarm among the residents of the city. Despite being aware of the possibilities of an earthquake, 309 people lost their lives, in large part because they lived in historic medieval buildings that had not been reinforced against earthquake shaking.

The average citizen of Addis Ababa, then, is much more likely to be aware of earthquakes than a citizen of Nairobi. But the earthquakes felt in Addis Ababa have not caused great economic or personal harm, and so the population treats them as more annoyances than critical concerns, ranked below famine, drought, and war. The engineers and seismologists of Addis Ababa are trying to raise awareness of earthquakes, but despite having more small earthquakes to illustrate the need for a response, upgrading building standards is an uphill struggle.

This section is dedicated to the late Father Pierre Gouin, who ran the seismograph station, did the basic research that has been used in this section, and raised awareness of earthquake hazards throughout the Rift Valleys, especially in Ethiopia.

PART III **Summary and recommendations**

28 **Where do we go from here?**

The study of earthquake time bombs around the world, including examples from the developed world (California, Cascadia, Japan, New Zealand, and Chile), concludes that (1) the outer layer of the Earth consists of moving tectonic plates that generate large, damaging earthquakes on their boundaries as well as within their interiors, (2) earthquake scientists are able to identify most, although not all, of the hazardous faults that are the sources of future earthquakes, (3) the rapid increase in population, particularly in megacities in the developing world, indicates that losses from the next earthquake will be many times larger than previous earthquakes on the same fault, as shown by the Haiti earthquake of 2010, and (4) earthquake scientists are as yet unable to predict the time, place, and magnitude of a future earthquake, even if the statistical probability of an earthquake on its source fault can be estimated.

Because earthquakes cannot yet be predicted, people, even in developed countries, are likely to take chances in their response to the threat of an earthquake. Real-estate developers may point out that the life of their development, a few hundred years, is small compared with the likelihood of the development being struck by an earthquake, a likelihood that may be measured in thousands of years. Homeowners may decide not to buy earthquake insurance, or they may not retrofit their homes and places of employment against earthquake damage. In addition, they will not demand that their state and local governments strengthen lifelines such as utility pipelines or transportation arteries like highway bridges and railroads, as discussed in the chapter on the next Cascadia earthquake.

Society tends to regard the study of earthquakes as academic research rather than a means to strengthen communities and economies against the effects of a large earthquake. Since scientists cannot answer the one question the public really wants to know – "When's the next Big One?" – decision makers gamble that the earthquake will not strike in their lifetimes, or, for political leaders, during their time in office.

Japan is the most prepared country on Earth against earthquakes, but even Japan underestimated the size of the magnitude 9 Tohoku-oki earthquake of 2011, resulting in the loss of nearly 16,000 lives, most from the tsunami that accompanied the earthquake.

The lack of preparedness is especially true in the developing world, where most of the Earth's population explosion is taking place. The location of earthquake faults and, in some cases, the probability of an earthquake on a fault or in a region is not enough incentive for government leaders to strengthen major cities against earthquakes, as demonstrated most recently by the 2010 Haiti earthquake. Decision makers either cannot afford to retrofit their cities, as was the case for Port-au-Prince, Haiti, or they may choose to ignore the problem, which is the case for Caracas, Kabul, and Tehran. The problem is made worse because the housing for recent migrants to developing-world cities is likely to be built by corrupt contractors, and inspection against construction flaws is of poor quality, even when the city has strong building codes. The people who are most likely to be killed in an earthquake are commonly recent arrivals looking for work, and they have no political clout with their government. Political leaders pay no price for failing to protect the lives of their citizens against an earthquake. In addition, the people at risk do not consider their own protection against earthquakes a high priority, assuming, if they think about earthquakes at all, that they will not be affected.

WHICH MEGACITIES ARE TAKING EARTHQUAKES SERIOUSLY, AND WHICH ARE NOT?

It is possible to subdivide cities at risk from catastrophic earthquakes into three groups: (1) cities in California, Japan, Chile, and New Zealand, where the earthquake hazard is being taken seriously;

(2) cities, mainly in the developing world, that are taking no meaningful action in preparation for the next earthquake; and (3) one megacity, Istanbul, that is committing major resources to reducing its exposure to earthquakes.

Cities in the first category (California, Japan, Chile, New Zealand) are all in the developed world, and all have been struck by destructive earthquakes in their recent past. The paradigm shift among local experts about the hazard from earthquakes has already taken place. In addition to studying earthquakes during the recorded history of cities in this category, local scientists have added to the historical record with geological, mainly paleoseismic and archaeological, evidence over thousands of years that increases the information available on the return times of earthquakes on the same fault. Outreach to the general population and to the media has made the case that the earthquake hazard is real, leading authorities to pass laws strengthening the communities under their responsibility against earthquakes, including retrofitting buildings and upgrading building codes and zoning laws that lead to safer, better-prepared citizens. When the earthquake hits, and these cities have been given enough time to prepare, their losses will be greatly reduced. People will die, and buildings will collapse, but the earthquake should not produce a catastrophe. Actions in cities in these regions are underway to protect their citizens even further by making them more resilient, although, as discussed for cities in these regions, there is still a major hazard, either due to pressure from property developers or from underestimations of size of the earthquake or accompanying tsunami. Even in many developed regions like the Pacific Northwest, the paradigm shift among the general public and government decision makers to adopt meaningful actions toward earthquake preparation has not taken place.

In this category of time bomb, there is a tendency to over-mitigate, and decision makers need to decide when the cost of mitigation is too high relative to the hazard being mitigated against. For example, Japanese planners raised costly seawalls on the northeastern coast of Honshu against a huge tsunami on the adjacent subduction zone, but it was not enough to prevent the loss of thousands of lives. The tsunami that overtopped the seawall was catastrophic, but the seawall built against a tsunami from a magnitude 8.4 earthquake would have worked for more than 1100 years. That argument will ring hollow for relatives of people drowned in the 2011 tsunami, obviously.

Other regions are candidates for this category: Italy, Greece, China, Australia, and other states in the United States, although Oregon, Washington, and British Columbia have not fully responded to the hazard from a subduction zone earthquake on the Cascadia subduction zone. All, including Cascadia, have major programs in earthquake hazard reduction, and scientists in those regions have made major contributions toward an understanding of earthquakes, not only in their own region but worldwide. China has the largest government-sponsored earthquake program in the world (China Earthquake Administration, with branches in every province), but despite major progress China still experienced the 2008 Wenchuan earthquake of magnitude 7.9 in which nearly 70,000 people lost their lives. Several cities in China have the potential for catastrophic earthquakes, including Beijing, the capital. Italy has the Istituto Nazionale di Geofisica e Volcanologia, which contains scientists who are world leaders in the field; Italy also hosts the headquarters of the Global Earthquake Model (GEM Foundation), discussed further below. Greece has a long history of earthquake investigations, although an earthquake in 1999 of magnitude 6 in a suburb of Athens activated a previously unknown fault and accounted for 143 deaths. Spain and Portugal are developing major earthquake programs, building on their early experience with the Lisbon earthquake of 1755. In the United States, Alaska, although it lacks a megacity, would probably be listed in the first category of prepared regions after its 1964 superquake of magnitude 9.2 in which 143 people were killed, including losses from the accompanying tsunami (an amazingly small number considering that the earthquake was the second largest ever recorded). Utah also has taken major measures against earthquakes on the Wasatch fault, since that fault extends close to major population centers, including Salt Lake City, which has a population of fewer than 200,000. Australia has megacities, but none are believed to be at major risk from earthquakes.

Decision makers in cities in the second category, if they think about earthquakes at all, play upon the fact that scientists are as yet unable to predict the time, place, and magnitude of a future damaging earthquake. For some earthquake time bombs in this category, like Port-au-Prince, Haiti, the local government does not have the resources to address the problem, and the local organizations that would have to strengthen the cities against earthquakes are in many places too corrupt to act responsibly.

The Istanbul metropolitan area is included as a time bomb, even though after the devastating earthquake of 1999 the Turkish authorities began major steps to increase Istanbul's resilience against earthquakes, to protect the population under their charge against great losses, as described in Chapter 15. An analysis of resilience not only includes response to earthquake damage, but a consideration of the long-term economic effect of the earthquake on the region as a whole. I still include Istanbul as a time bomb because it is a work in progress. Some of my Turkish colleagues believe that the corruption problem in Turkey is too large to overcome. The success of Turkey's actions will not be known until the next expected earthquake strikes the city in the near future. However, I hope that Istanbul will provide a way forward for other time bomb cities. The Istanbul chapter describes the difficult task undertaken in that city, with the outcome still uncertain. But there is hope.

STEPS NOW BEING TAKEN

The chapter "Population explosion and increased risk to megacities" explains why future earthquakes on the same fault will be much more devastating than previous earthquakes because of the dramatic increase in population of megacities, particularly those in the developing world. The chapter "When's the next big one?" covers plans presently underway to address the present earthquake problem to society.The Global Earthquake Model (GEM) is a public–private partnership that started its planning in 2006 through the Organisation for Economic Co-operation and Development (OECD). This organization, headquartered in Paris, includes thirty-four developed countries around the world, together with working relations with other countries in the developed and developing world. GEM is creating software to aid countries at risk of earthquakes (see their website at www.globalquakemodel.org). GEM began developing databases, models, guidelines, and tools in 2009 (Christofferson *et al.*, 2015); their publicly accessible OpenQuake Platform launched in late 2014. The earthquake insurance industry and major engineering consulting firms are particularly interested in GEM because it proposes to tell them what their global exposure to earthquake risk is, what would be necessary to reduce their exposure, and what the impact of retrofit and relocations would be on earthquake losses and economic recovery after the earthquake. This interest has led to major funding for GEM by the insurance and engineering-consulting industries: more than 10 million euros.

The *Global Earthquake Model (GEM)* is a public–private partnership that started its planning in 2006 through the Organisation for Economic Co-operation and Development (OECD). This organization, headquartered in Paris, includes 34 developed countries around the world, together with working relations with other countries in the developed and developing world. GEM is creating software to aid countries at risk of earthquakes (see their website at www.globalquakemodel.org). GEM began developing databases, models, guidelines, and tools in 2009; their publicly accessible Open-Quake Platform launched in late 2014. The earthquake insurance industry and major engineering consulting firms are particularly interested in GEM because it proposes to tell them what their global exposure to earthquake risk is, what would be necessary to reduce their exposure, and what the impact of retrofit and relocations would be on earthquake losses and economic recovery after the earthquake. This interest has led to major funding for GEM by the insurance and engineering-consulting industries: more than €10 million.

GEM's scientific framework is subdivided into three main modules. The *hazard module* will provide the probability of the occurrence of earthquakes and of strong shaking at any given location on Earth. The *risk module* will focus on damage and direct losses, including fatalities, injuries, and the cost of repairs. The *socio-economic vulnerability and resilience module* focuses on ways to estimate the vulnerability of society and the economy and the capacity to cope with earthquake events. The integrated seismic risk obtained through these three modules will take into account the cost–benefit analyses of possible risk mitigation actions to be taken: strengthen buildings, engage the public in emergency preparation, and transfer risk within the insurance industry and various levels of government. This final result is the most difficult to put into practice, as illustrated by the slow response of decision makers to recent resilience surveys in Oregon and Washington.

The GEM Foundation is headquartered in Pavia, Italy, hosted by the European Centre for Training and Research in Earthquake Engineering (EUCENTRE), under the auspices of the Italian government. In addition to developing uniform global models and datasets, regional programs have also been initiated worldwide, with a view to engage the local scientific communities in the development of regional national and sub-national hazard and risk models. A second five-year plan extends the GEM project to 2019.

Another option has been proposed by Max Wyss, director of the World Agency of Planetary Monitoring and Earthquake Risk Reduction in Geneva, Switzerland. Wyss looked at the 12 earthquakes in the past ten years that caused more than 1000 deaths each, including the three most deadly earthquakes: Kashmir 2005, Wenchuan, China, 2008, and Haiti 2010. The number of estimated fatalities predicted using only the probabilistic method underestimated the number actually killed by 160 to 1. In addition, the probabilistic estimates are focused on the earthquake itself and not on the people at risk and the condition of the buildings that might collapse. Wyss advocates using the deterministic method, which estimates the magnitude of the worst-case earthquake that would strike faults affecting the region. This method is well known for evaluating the earthquake hazard to critical facilities like power plants and dams, although here it would apply to metropolitan areas and would take into account the reinforcement against earthquakes (for example, the expected magnitude 9 Cascadia earthquake) of buildings of various strengths. A deterministic analysis will probably have the effect of raising the stakes of alarm for the next earthquake and increasing the maximum considered earthquake above magnitude 9. Would this be enough for Northwesterners and their governing officials to take action?

In addition, as discussed in Chapter 8 on Cascadia, the states of Oregon and Washington, authorized by their respective state legislatures, have completed resilience surveys with similar goals to GEM. These surveys focused on the Cascadia subduction zone, which is expected to generate an earthquake of magnitude 9, the same size as the 2011 Tohoku-oki, Japan, earthquake. The Oregon survey, coordinated by the Oregon Seismic Safety Policy Advisory Commission (OSSPAC), estimated the cost of doing nothing or of taking only token steps. The survey, prepared by groups of experts in fields ranging from earthquake engineering to planning, transportation, and utility lifelines, estimates how many businesses will be destroyed, how many lives will be lost, and how long it will take to get lifelines back in service. In some cases, this would be as long as two years. For businesses on the Oregon coast, extended isolation from their markets would mean that businesses would have to close or relocate, taking the jobs and tax base with them. The economy of Oregon could be diminished for as long as a generation.

Although the Oregon resilience survey was published while the Oregon state legislature was in session, the print and television media failed to see the implications of the survey's results, and so there was very little impact on the general public. I did my best in an interview with Associated Press, but the story that was produced was disappointing and failed to engage the public. The legislature had many pressing problems, and they did not include earthquakes as a high priority.

However, the long-term implications of the survey are slowly making themselves known. I was interviewed a second time by Associated Press. People involved in the OSSPAC survey have been invited to testify before state legislative committees, and these are being covered by the media, although as yet without any sense of urgency. Scott Ashford, Dean of Engineering at Oregon State University and an earthquake engineering specialist, has been authorized by the Oregon legislature to chair a Governor's Task Force on responding to the resilience survey. The Portland School District, the largest in Oregon, passed a bond issue to improve its schools, and funds from that bond issue will be used for seismic upgrades of school buildings, many of which were built prior to the development of building codes that addressed earthquake hazard. Similarly, the Oregon Department of Transportation is responding to the OSSPAC committee's conclusions about obsolete bridges, although the legislature has not appropriated enough money to repair Oregon's obsolete bridges. Obviously, the steps necessary to strengthen bridges and utility lifelines cannot be done in a single year, but if a plan is made to do it over a 50-year period, then decision makers reduce their gamble with people's lives and safety by hoping that the next Cascadia earthquake will not strike during that period. Even that is a gamble, but it is more responsible than doing nothing or taking only token steps. A price tag of $100 million per year has been advanced, in hopes that the strengthening will be completed before the earthquake strikes. Who will pay for this?

The state of California, with its earthquake history, has been more responsive to earthquake preparation than elsewhere in the United States. Mandatory retrofit programs are underway, and that part of the Cascadia subduction zone in California is better prepared than coastal cities farther north in Oregon and Washington. The San Francisco Bay Area is engaged in a ten-year mandatory seismic retrofit program that should greatly strengthen the region against the next earthquake.

Los Angeles was slower to take up the challenge, but the Great California ShakeOut, organized by the Southern California Earthquake Center, involved millions of Californians in responding to an earthquake of magnitude 7.8 on the southern San Andreas fault, which has not had a great earthquake in more than 300 years. The City of Los Angeles then used this future San Andreas earthquake to plan its response to that earthquake in terms of retrofitting unreinforced-masonry buildings and nonductile concrete buildings against earthquakes, upgrading water facilities so that the city could function and could fight fires after the earthquake, and upgrading its telecommunications networks, including Internet connections and cell phones. The city contracted with the USGS, to provide seismologist Lucile Jones as a source of technical support for its planned upgrades.

Istanbul, listed as an earthquake time bomb in this book, is under warning from three independent studies of another earthquake in the near future. This led Istanbul's governing bodies to solicit recommendations from Turkish experts from four major universities on how to deal with an earthquake as large as the 1999 Marmara earthquake. As summarized by one of these experts, Professor Mustafa Erdik of Bogaziçi University, those recommendations gave rise to successful applications for loans from the World Bank and other lending institutions to strengthen Istanbul against the forecasted earthquake, as discussed in Chapter 15 on Istanbul. The authorities have adopted a retrofit program over the next 20 years that will cost $200 billion. Upgrading of hospitals and schools is already underway. Engineers are being trained to conduct assessments of the earthquake safety of new buildings, attempting to resolve the corruption problem in the construction industry that resulted in the tragic loss of so many lives in past earthquakes.

The general public is now involved in earthquake safety in homes and workplaces. This is still a work in progress, but my hope is that at the end of this project, Istanbul can be placed in the same category as San Francisco, Tokyo, Santiago, and Wellington. These cities are all close to earthquake faults, but their preparations should greatly reduce the losses when the next earthquake strikes.

Why Istanbul? Two reasons stand out. First, this part of Turkey had experienced two major earthquakes in 1999, with thousands of lives lost and billions of lost value in destroyed buildings. This was followed by several probabilistic earthquake forecasts, all of which were very specific in pointing out the immediate hazard to Istanbul, more specific than for any other megacity on Earth. Because the economy of the Istanbul metropolitan area is a large part of the economy of the entire country, the decision makers rose to the occasion, and a 20-year plan is underway to retrofit buildings, starting with hospitals and schools and extending to residential housing. This social experiment will be tested in the next earthquake, which the probability forecasts state will be in the very near future.

The main question for the cities covered by resilience surveys is the same: who pays?

The difference between Istanbul and other earthquake time bombs, then, was the forecast. Tehran, Kabul, Caracas, Damascus, Aleppo, Beirut, and Athens are at risk from earthquakes, but the probabilistic hazard forecast is not as detailed and specific as it is for Istanbul.

One other region has a probabilistic forecast that is as specific as that for Istanbul: Cascadia, where I live. The probabilistic forecast for a subduction zone earthquake on the southern part of the subduction zone in southern Oregon and northern California is 37% for a magnitude 8–8.4 earthquake in the next 30 years. The chance of a magnitude 9 earthquake over the entire subduction zone from northern California to Vancouver Island is 10% in the next 50 years. These forecasts are based on a history of Cascadia earthquakes over the past 10,000 years, the longest such record on any subduction zone on Earth. Washington and Oregon have conducted resilience surveys for the subduction zone, with stark implications for the time that it could take for coastal Oregon and Washington to recover from a subduction zone earthquake. However, the resilience surveys and the probabilistic forecasts have not yet led to a call for action comparable to that taken for Istanbul by decision makers in either state or by the general public. In this respect, Cascadia ranks closer to Tehran than it does to San Francisco.

For other megacities in the developing world, the outlook is grim. Port-au-Prince has yet to recover from its devastating earthquake of 2010, in part because help from the outside world was inadequate to deal with the problem of reconstruction after the last earthquake, let alone strengthen the city against the next earthquake. Kabul has received large sums of money from the developed world, especially NATO countries, to deal with the Taliban insurgency, but very little of this money is earmarked for strengthening the city against earthquakes, even though a repeat of the 1505 earthquake could kill as many people as all the wars in Afghanistan for the past 40 years. Because of their oil revenues, Caracas and Tehran can afford to adopt earthquake upgrades comparable to those in Istanbul, but as yet there is no evidence that decision makers have made these upgrades a high priority. Ecuador, another major oil-producing country, is taking the earthquake threat more seriously. Manila, Yangon (Myanmar), and Nairobi also show no concerted effort to upgrade their low-income residential areas against earthquakes.

An American seismologist, Dr. Brian Tucker, founded a nonprofit firm called GeoHazards International (GHI) to work with local governments to improve the safety of their cities. The first project was Kathmandu, Nepal, and now people trained by GHI and a local organization connected with Kathmandu city government have gone out to other countries to assist in retrofit projects after an earthquake. A second project focused on schools in Quito, Ecuador, and led to upgrades in building codes in Quito and education on strengthening buildings against earthquakes, again involving local government, local universities, and industry. Both Nepal and Ecuador have local experts who provide leadership and technical support. In addition, the projects have gained the support of international corporations such as Oyo Corporation in Quito, Ecuador.

As described above, the Southern California Earthquake Center developed a plan involving the general public, called the *Great California ShakeOut*. Millions of people are involved, and the plan is widely covered by the print and television media. The ShakeOut has been adopted by communities around the world. Public engagement is critical to its success.

A FINAL WORD

This book is a progress report, and changes will take place in part driven by the next damaging urban earthquake. It is possible, perhaps even likely, that the next urban earthquake will strike a city not listed here as an earthquake time bomb. This speaks to our lack of knowledge about the regions most likely to generate an earthquake in the next few decades. The Christchurch, New Zealand, earthquakes are a reminder of how much we have to learn to do a better job of convincing the public to take action before their city is struck by an earthquake.

The most positive development in the past few years is the development of resilience surveys, in which the focus is not just on immediate recovery from the earthquake but also the long-term economic effects of the earthquake, including loss of population and loss of business, from which a time bomb may not recover for as long as a generation. When this is fully appreciated, expenditures to become a resilient community should increase to meet the challenge. Let's hope the earthquake waits long enough.

References

Ambraseys, N. N., 1991, The Rukwa earthquake of 13 December 1910 in East Africa. *Terra Nova*, v. 3, pp. 202–211.

Ambraseys, N., 2009, *Earthquakes in the Mediterranean and Middle East.* Cambridge, Cambridge University Press.

Ambraseys, N. and Bilham, R., 2011, Corruption kills. *Nature*, v. 469, pp. 153–155.

Ambraseys, N. and Melville, C., 1982, *A history of Persian earthquakes.* Cambridge, Cambridge University Press.

Andreani, L., Le Pichon, X., Rangin, C., and Martinez-Reyes, J., 2008, The southern Mexico block: Main boundaries and new estimation for its Quaternary motion: *Bull. Soc. Geologique de France*, v. 117, pp. 209–223.

Atwater, B. F., Cisternas, V., M., Bourgeois, J., Dudley, W. C., Hendley II, J. W., and Stauffer, P. H., 1999, *Surviving a tsunami: Lessons from Chile, Hawaii, and Japan.* US Geological Survey, Circular 1187 (updated in 2005).

Atwater, B. F., Satoko, M.-R., Satake, K., Tsuji, Y., Kazue, U., and Yamaguchi, D. K., 2005, *The orphan tsunami of 1700: Japanese clues to a parent earthquake in North America.* Reston, VA, USGS and University of Washington Press.

Audemard, F. A., Machette, M. N., Cox, J. W., Dart, R. L., and Haller, K. M., 2000, *Map and database of Quaternary faults in Venezuela and its offshore regions.* USGS Open-File Report OFR 00-0018.

Bakun, W. H. and Wentworth, C. M., 1997, Estimating earthquake location and magnitude from seismic intensity data. *Seismol. Soc. Am., Bull.*, v. 87, pp. 1502–1521.

Berberian, M. and Yeats, R. S., in press, *Tehran, an earthquake time bomb.* Geological Society of America.

Bischke, R. E., Suppe, J., and del Pilar, R., 1990, A new branch of the Philippine fault system as observed from aeromagnetic and seismic data. *Tectonophysics*, v. 183, pp. 243–264.

Bolt, B. A., 2004, *Earthquakes*, 5th edn. New York, W.H. Freeman and Co.

Cascadia Region Earthquake Workgroup (CREW), 2013: *Cascadia Subduction Zone Earthquakes: a magnitude 9.0 earthquake scenario, update 2013.*

Christofferson, A., *et al.*, 2015, Development of the Global Earthquake Model's neotectonic fault database: Natural Hazards, v. 79, no. 1, p. 111–135.

City of Los Angeles, 2014, *Resilience by design.* http://lamayor.org/earthquakes

Cowan, H. A., Prentice, C., Pantosti, D., *et al.*, 2002, Late Holocene earthquakes on the Aeropuerto fault, Managua, Nicaragua. *Seismological Soc. America Bull.*, v. 92, pp. 1694–1707.

Draper, R. and Girard, G., 2010, Kings of controversy: The search for King David. New discoveries in the Holy Land. *National Geographic*, v. 218, no. 6, pp. 66–91.

Erdik, M., 2013, Earthquake risk in Turkey. *Science*, v. 341, p. 724–725.

Espinosa, A. F. (ed.), 1976, *The Guatemalan earthquake of February 4, 1976: A preliminary report.* USGS Professional Paper 1002, 90 p.

Finkel, M. and Paley, M., 2013, Stranded on the roof of the world. *National Geographic*, v. 223, no. 2, pp. 84–111.

Fonseca, J. D., 2004, *O Terremoto de Lisboa/ The Lisbon Earthquake.* Lisbon, Argumentum Press.

Geschwind, C.-H., 1996, *Earthquakes and their interpretation: The campaign for seismic safety in California, 1906–1933*: Johns Hopkins University PhD dissertation, 256 p.

Geschwind, C.-H., 2001, *California earthquakes: Science, risk, and the politics of hazard mitigation, 1906–1977*: Baltimore, MD, Johns Hopkins University Press.

Gouin, P., 1979, *Earthquake history of Ethiopia and the Horn of Africa.* Ottawa, International Development Research Centre.

Gutscher, M. A., Baptista, M. A., and Miranda, J. M., 2006, The Gibraltar Arc seismogenic zone (part 2): Constraints on a shallow east-dipping fault plane source for the 1755 Lisbon earthquake provided by tsunami modeling and seismic intensity. *Tectonophysics*, v. 426, pp. 153–166.

Hall, J. K., 1994, Digital shaded-relief map of Israel and environs 1:500,000. Israel Geological Survey and S. Marco, Tel Aviv University.

Holmes, A., 1978, *Principles of physical geology.* New York: John Wiley & Sons.

Jones, L. M., Bernknopf, R., Cox, D., *et al.*, 2008, *The ShakeOut Scenario.* USGS Open File Report 2008/1150 and California Geological Survey Preliminary Report 25. http://pubs.usgs.gov/of2008/1150 and http://conservation.ca.gov/cgs.

Jones, L. M., 2015, Resilience by design: Bringing science to policy makers. *Seismol. Res. Lett.*, v. 86, pp. 294–300.

Jordan, T. H., 2013, Lessons of L'Aquila for operational earthquake forecasting. *Seismol. Res. Lett.*, v. 84, no. 1, pp. 4–7.

Jordan, T. H., Chen, Y.-T., Gasparini, P., *et al.*, 2011, Operational earthquake forecasting: State of knowledge and guidelines for implementation. Final Report of the International Commission on Earthquake Forecasting for Civil Protection. *Ann. Geophys.*, v. 54, pp. 315–391.

Jordan, T. H., Marzocchi, W., Michael, A. J., and Gerstenberger, M. C., 2014, Operational earthquake forecasting can enhance earthquake preparation. *Seismol. Res. Lett.*, v. 85, no. 5, pp. 955–959.

Kondo, H., Awata, Y., Emre, O., et al., 2005, Slip distribution, fault geometry, and fault segmentation of the 1944 Bolu-Gerede earthquake rupture, North Anatolian fault, Turkey. *Seismol. Soc. Am. Bull.*, v. 95, pp. 1234–1249.

Kovach, R. L., 2004, *Early earthquakes of the Americas*. New York, Cambridge University Press, 268 p.

LaFeber, W., 1983, *Inevitable revolutions*. New York, W. W. Norton & Co.

Lawrence, R. E., Khan, S. H., and Nakata, T., 1992, Chaman fault, Pakistan–Afghanistan. *Annales Tectonicae Supp.*, v. 6, pp. 196–223.

Lin, W., Conin, M., Moore, J. C., Chester, F. M., et al., 2013, Stress state in the largest displacement area of the 2011 Tohoku-oki earthquake. *Science*, v. 339, pp. 687–690.

Lyell, C., 1830–1833, *Principles of Geology*, 3 volumes. London, John Murray, 12th edition published in 1875.

McCullough, D., 1977, *The path between the seas: The creation of the Panama Canal, 1870–1914*. New York, Simon & Schuster.

Metzger, E. P. (n.d.), Sea-floor spreading teacher's guide. www.ucmp.berkeley.edu

Mugnier, J.-L., Gajurel, A., Huyghe, F., Jayangondaperumal, R., Jouanne, F., and Upreti, B., 2013, Structural interpretation of the great earthquakes of the last millennium in the central Himalaya. *Earth Sci. Rev.*, v. 127, pp. 30–47.

Neev, D. and Emery, K. O., 1995, *The destruction of Sodom, Gomorrah, and Jericho*. New York, Oxford University Press.

Olson, R. S., Podesta, B., and Nigg, J. M., 1989, *The politics of earthquake prediction*. Princeton, NJ, Princeton University Press.

Oregon Seismic Safety Policy Advisory Commission (OSSPAC), 2013, The Oregon resilience plan: Reducing risk and improving recovery for the next Cascadia earthquake and tsunami. www.oregon.gov/OMD/OEM//osspac/docs/oregon-Resilience_Plan_Final.pdf.

Plafker, G., 1977, Color slides showing geologic effects. USGS Open-File Report 77-165.

Population Reference Bureau, 1994, *World population: Toward the next century*. Washington, DC, Population Reference Bureau.

Porter, K., Jones, L., Cox, D., et al., 2011, The ShakeOut Scenario: A hypothetical Mw7.8 earthquake on the southern San Andreas fault. *Earthquake Spectra*, v. 27, pp. 239–261.

Richter, C. F., 1958, *Elementary seismology*. San Francisco, CA, W. H. Freeman & Co.

Rosen, J., 2015, Los Angeles gets serious about preparing for the "Big One:" EOS. *Earth and Space Science News*, v. 96, no. 4, pp. 18–23.

Schulz, K., 2015, The really big one. *The New Yorker*, July 20, pp. 52–9.

Searle, M., Chung, S.-L., and Lo, C.-H., 2010, Geological offsets and age constraints along the northern Dead Sea fault, Syria. *Jour. Geol. Soc. London*, v. 167, pp. 1001–1008.

Sébrier, M., Mercier, J. L., Mégard, F., Laubacher, G., and Carey Gailhardis, D., 1985, Quaternary normal and reverse faulting and the state of stress in the central Andes of south Peru. *Tectonics*, v. 4, pp. 739–780.

Spence, W., Herrmann, R. B., Johnston, A. C., and Reagor, G., 1993, Responses to Iben Browning's prediction of a 1990 New Madrid, Missouri, earthquake. USGS Circular 1083.

Stein, S., 2014, How much natural hazard mitigation is enough? *Earth*, v. 59, no. 11–12, pp. 14–16.

Sugiyama, Y., 1994, Neotectonics of Southwest Japan due to the right-oblique subduction of the Philippine Sea Plate. *Geofísica Internacional*, v. 33, pp. 53–76.

Thompson, J., 2011, *Cascadia's fault*. Toronto, HarperCollins.

Transparency International, 2014, Corruption Perceptions Index. www.transparency.org/policy_research/surveys_indices/cpi/2014.indetail4 (updated yearly).

Tsutsumi, H. and Sato, T., 2009, Tectonic geomorphology of the southernmost Sagaing fault and surface rupture associated with the May 1930 Pegu (Bago) earthquake, Myanmar. *Seismol. Soc. Am. Bull.*, v. 99, pp. 2155–2168.

Voltaire, 1758, *Candide*. Paris, Sirène.

Wallace, R. E. (ed.), 1990, The San Andreas fault system, California. USGS Professional Paper 1515.

Wang, K. (2006). Predicting the 1975 Haicheng Earthquake. *Seismol. Soc. Am. Bull.*, v. 96, pp. 757–795.

Wang, K. and Rogers, G. C., 2014, Earthquake preparedness should not fluctuate on a daily or weekly basis. *Seismol. Res. Lett.*, v. 85, no. 3, pp. 569–571.

Wegener, A. L., 1915, *The origin of continents and oceans (Die Enstehung der Kontinente und Oceane)* (English edition by Dover Pubs., Inc., 1966).

Wilson, J. T., 1965, A new class of faults and their bearing on continental drift. *Nature*, v. 207, pp. 343–347.

Winkler, W., Villagomez, D., Spikings, R., *et al.*, 2005, The Chota basin and its significance for the inception and tectonic setting of the inter-Andean depression in Ecuador. *J. South Am. Earth Sci.*, v. 19, pp. 5–19.

Yeats, R. S., 2004, *Living with earthquakes in the Pacific Northwest*, 2nd edn. Corvallis, OR, Oregon State University Press. http://oregonstate.edu/instruct/oer/earthquake.pdf.

Yeats, R. S., 2012, *Active Faults of the World*. Cambridge, Cambridge University Press.

Yeats, R.S., 2014, Partnering with Cuba: earthquake hazards. *Science*, v. 345, p. 278.

Zhang, Y. Q., Mercier, J. L., and Vergely, P., 1998, Extension in the graben systems around the Ordos (China) and its contribution to the extrusion tectonics of south China with respect to Gobi-Mongolia. *Tectonophysics*, v. 285, pp. 41–75.

Index